Directed Reading A

Section: Human Reproduction

THE MALE REPRODUCTIVE SYSTEM

Match the correct description with the correct term. Write the letter in the space provided.

______ **1.** This produces fluid that mixes with sperm.

______ **2.** This makes sperm and testosterone.

______ **3.** Semen travels through this tube.

______ **4.** This outside organ puts semen into a female.

a. urethra
b. testis
c. penis
d. prostate gland

Match the correct description with the correct term. Write the letter in the space provided.

______ **5.** male sex hormone

______ **6.** place where sperm are stored

______ **7.** tube that leads to the prostate gland

______ **8.** mixture of sperm and fluid

a. vas deferens
b. semen
c. testosterone
d. epididymis

Delivery of Sperm

Write the letter of the correct answer in the space provided.

______ **9.** Which of the following statements is NOT true?
  a. Fertilization occurs when sperm enter an egg.
  b. Fertilization can only occur after the male ejaculates.
  c. Few sperm are necessary for fertilization to occur.
  d. Fertilization can occur without the male ejaculating.

THE FEMALE REPRODUCTIVE SYSTEM

______ **10.** What are the female organs that make eggs?
  a. the fallopian tubes
  b. the uterus
  c. the ovaries
  d. the vagina

Directed Reading A *continued*

_______ **11.** Which of the following are female sex hormones?
    a. testosterone and estrogen
    b. chromosome and testosterone
    c. estrogen and progesterone
    d. ovulation and estrogen

The Egg's Journey

Match the correct description with the correct term. Write the letter in the space provided.

_______ **12.** place where eggs are usually fertilized

_______ **13.** when an egg is released from the ovary

_______ **14.** place where a fertilized egg develops

_______ **15.** passage through which babies come out during birth

a. uterus

b. fallopian tube

c. vagina

d. ovulation

Menstrual Cycle

Match the correct description with the correct term. Write the letter in the space provided.

_______ **16.** when the uterus sheds blood and tissue

_______ **17.** a complete menstrual cycle

_______ **18.** when ovulation happens

a. 14th day of menstrual cycle

b. day 1 of menstrual cycle

c. about 28 days

FERTILIZATION

Write the letter of the correct answer in the space provided.

_______ **19.** How many copies of each chromosome are in a fertilized egg?
    a. one
    b. two
    c. three
    d. four

Directed Reading A *continued*

MULTIPLE BIRTHS

Match the correct description with the correct term. Write the letter in the space provided.

______ **20.** having more than one baby at a time

______ **21.** twins that have the exact same genes

______ **22.** twins that don't have the exact same genes

a. identical twins
b. multiple birth
c. fraternal twins

REPRODUCTIVE SYSTEM PROBLEMS

Sexually Transmitted Diseases (STDs)

Write the letter of the correct answer in the space provided.

______ **23.** How does a person get a sexually transmitted disease?
    a. from coughing
    b. from sexual contact
    c. from dirty bathrooms
    d. from shaking hands

______ **24.** Which one of the following is an STD, or sexually transmitted disease?
    a. AIDS
    b. the flu
    c. a cold
    d. cancer

______ **25.** How many new hepatitis B cases occur in the United States each year?
    a. none
    b. 100,000
    c. 120,000
    d. 140,000

Cancer

______ **26.** Cancer is caused by the uncontrolled growth of what?
    a. eggs
    b. zygotes
    c. cells
    d. the uterus

______ **27.** What is a common reproductive cancer of men?
    a. prostate cancer
    b. penis cancer
    c. liver cancer
    d. cancer of the cervix

Directed Reading A *continued*

_______ **28.** What is a common reproductive cancer of women?
    a. prostate cancer
    b. penis cancer
    c. liver cancer
    d. cancer of the cervix

Infertility

_______ **29.** Infertile couples cannot do what?
    a. produce sperm
    b. get STDs
    c. have children
    d. get cancer

_______ **30.** What is one cause of infertility in men?
    a. liver disease
    b. too few offspring
    c. abnormal ovulation
    d. few healthy sperm

Directed Reading A

Section: Growth and Development
FROM FERTILIZATION TO EMBRYO

Write the letter of the correct answer in the space provided.

______ **1.** Where does the sperm usually fertilize the egg?
- **a.** in the uterus
- **b.** in a fallopian tube
- **c.** in a membrane
- **d.** in a nucleus

______ **2.** What is an embryo?
- **a.** a newborn baby through two years of life
- **b.** an unfertilized egg
- **c.** a fertilized egg through week 10 of pregnancy
- **d.** a million sperm

______ **3.** What is it called when the embryo attaches itself to the uterus?
- **a.** fertilization
- **b.** implantation
- **c.** multiple birth
- **d.** menstrual cycle

FROM EMBRYO TO FETUS

______ **4.** Through what organ does the mother nourish the developing embryo?
- **a.** the uterus
- **b.** the placenta
- **c.** the fallopian tube
- **d.** the cervix

Weeks 1 and 2

______ **5.** When do doctors start counting the time of a woman's pregnancy?
- **a.** from the first day of fertilization
- **b.** from the first day of her last menstrual period
- **c.** from the first day the embryo is implanted
- **d.** from the time sperm come into the uterus

Weeks 3 and 4

______ **6.** When does fertilization take place?
    a. the end of day 1
    b. the end of week 1
    c. the end of week 2
    d. when the embryo is implanted

______ **7.** Which of the following begins during the first 4 weeks of pregnancy?
    a. The embryo grows fingernails.
    b. The embryo can kick its feet.
    c. The embryo can see light.
    d. The embryo's blood cells begin to form.

Weeks 5 to 8

______ **8.** What connects the embryo to the placenta?
    a. the amnion
    b. the umbilical cord
    c. the fallopian tube
    d. the spinal cord

______ **9.** During weeks 5 to 8, what part of the embryo grows quickly?
    a. its arms
    b. its muscles
    c. its brain
    d. its taste buds

Weeks 9 to 16

______ **10.** What is the unborn child called after week 10?
    a. a fetus
    b. an embryo
    c. an infant
    d. a zygote

______ **11.** What happens to the fetus during weeks 9 to 16?
    a. It responds to light.
    b. The umbilical cord forms.
    c. It can hear sounds.
    d. It begins to move.

Directed Reading A *continued*

Weeks 17 to 24

_______ **12.** By week 18, what can the fetus respond to?
 a. light
 b. language
 c. sound
 d. colors

Weeks 25 to 36

_______ **13.** By week 32, brain activity shows that the fetus may respond to what?
 a. odors
 b. light
 c. touch
 d. temperature

BIRTH

_______ **14.** What are the contractions a mother feels when giving birth?
 a. full-term
 b. placenta
 c. labor
 d. uterus

_______ **15.** A baby comes out of what part of a mother's body?
 a. her vagina
 b. her placenta
 c. her umbilicus
 d. her cervix

FROM BIRTH TO DEATH
Infancy and Childhood

_______ **16.** A child is called an infant until it reaches what age?
 a. 1 year old
 b. 2 years old
 c. 3 years old
 d. 5 years old

_______ **17.** What is one change you experience during childhood?
 a. You get baby teeth.
 b. You grow hair.
 c. You learn how to walk.
 d. You get permanent teeth.

Directed Reading A *continued*

Adolescence

_______ **18.** What happens to you during puberty?
    a. You get permanent teeth.
    b. Your nervous system develops.
    c. Your reproductive system matures.
    d. Your muscles become coordinated.

Adulthood

_______ **19.** What reaches its peak when you are a young adult?
    a. your aging process
    b. your physical development
    c. your body fat
    d. your wealth

_______ **20.** What is a common sign of aging in older adults?
    a. greater flexibility
    b. blindness
    c. graying of hair
    d. inability to walk

Directed Reading B

Section: Human Reproduction

THE MALE REPRODUCTIVE SYSTEM

Match the correct description with the correct term. Write the letter in the space provided.

_______ **1.** place where male's sperm are produced

_______ **2.** male hormone that regulates the development of sperm

_______ **3.** mixture of sperm and other fluids

_______ **4.** place where sperm mix with fluid

_______ **5.** tube that stores sperm

_______ **6.** tube through which semen is carried out of the penis

_______ **7.** male's exterior sex organ

_______ **8.** when a sperm enters an egg

a. testes
b. urethra
c. epididymis
d. semen
e. vas deferens
f. testosterone
g. penis
h. fertilization

THE FEMALE REPRODUCTIVE SYSTEM

_______ **9.** A female's eggs and her sex hormones are produced in her
    a. zygote.
    b. estrogen.
    c. fallopian tubes.
    d. ovaries.

_______ **10.** An egg travels into a fallopian tube during
    a. menstruation.
    b. ovulation.
    c. fertilization.
    d. pregnancy.

_______ **11.** Fallopian tubes connect the ovaries with the
    a. uterus.
    b. zygote.
    c. vagina.
    d. eggs.

Directed Reading B *continued*

_______ **12.** A zygote develops in the female's
    a. ovary.
    b. vagina.
    c. uterus.
    d. placenta.

_______ **13.** The canal between the outside of the body and the uterus is the
    a. zygote.
    b. placenta.
    c. vagina.
    d. fallopian tubes.

14. When does a woman begin menstruating? When does menstruation usually end?

____

____

15. On the first day of the menstrual cycle, what does the uterus discharge?

____

16. What occurs on about the 14th day of the menstrual cycle?

____

17. About how long does a complete menstrual cycle last?

____

FERTILIZATION

18. When must sperm be present in the female's reproductive system for fertilization to occur?

____

MULTIPLE BIRTHS

19. Children who are born at the same time and who share the exact same genes are called ____________________.

20. Of all twin births, identical twins make up about ____________________.

21. Children who are born at the same time but who can look very different from each other are called ____________________.

22. A rare type of multiple birth occurring when a mother gives birth to five babies is called ____________________.

Directed Reading B *continued*

REPRODUCTIVE SYSTEM PROBLEMS

_____ **23.** Which one of the following is a sexually transmitted disease (STD)?
    a. influenza
    b. tuberculosis
    c. herpes
    d. smallpox

_____ **24.** How can an infected person pass on an STD to another person?
    a. through sneezing
    b. through casual contact
    c. by shaking hands
    d. through sexual contact

_____ **25.** Hepatitis B is an STD that mainly affects which organ?
    a. the penis
    b. the heart
    c. the liver
    d. the uterus

_____ **26.** AIDS is an STD that is caused by
    a. herpes.   **c.** the immune system.
    b. a virus.   **d.** poor hygiene.

_____ **27.** Which one of the following is a common reproductive system cancer among men?
    a. cancer of the prostate
    b. cancer of the liver
    c. cancer of the vas deferens
    d. cancer of the blood

_____ **28.** Which one of the following is a common reproductive system cancer among women?
    a. cancer of the vagina
    b. cancer of the breast
    c. cancer of the neck
    d. cancer of the fallopian tubes

_____ **29.** What percentage of couples in the United States have difficulty producing offspring?
    a. 5   **c.** 15
    b. 10   **d.** 20

30. How often do STDs cause infertility in men as compared to women?

____

____

Directed Reading B

Section: Growth and Development
FROM FERTILIZATION TO EMBRYO

______ **1.** Ordinarily, how many sperm fertilize a female's egg?
  a. several million
  b. a few hundred
  c. one
  d. six to ten

______ **2.** An egg is fertilized after the union of the egg's and sperm's
  a. membranes.
  b. nuclei.
  c. semen.
  d. zygote.

______ **3.** About 5 days after fertilization, the egg arrives at the
  a. fallopian tubes.
  b. vagina.
  c. cervix.
  d. uterus.

______ **4.** Which one of the following is a characteristic of an embryo?
  a. It is a tiny ball of cells.
  b. It is not yet fertilized.
  c. It develops in the vagina.
  d. It is less than 1 day old.

FROM EMBRYO TO FETUS

______ **5.** Which of the following is a characteristic of the placenta?
  a. It's an exchange organ between mother and embryo.
  b. It contains very few blood vessels.
  c. It becomes an organ in the developing embryo.
  d. It contracts to help the mother give birth.

______ **6.** The child developing inside the mother's uterus is connected to the placenta by a tube called
  a. the vagina.
  b. an embryo.
  c. the amnion.
  d. the umbilical cord.

Directed Reading B *continued*

_______ **7.** After 10 weeks, a developing embryo is called a(n)
   a. amnion.
   b. fetus.
   c. zygote.
   d. placenta.

Match the correct description with the correct term. Write the letter in the space provided.

_______ **8.** The embryo's blood cells begin to form.

_______ **9.** The spinal cord and regions of the brain begin to develop.

_______ **10.** The fetus's eyes can respond to light.

_______ **11.** The fetus doubles in size and begins to move.

_______ **12.** The fetus is able to hear sounds.

a. weeks 2–4

b. weeks 5–8

c. weeks 9–16

d. weeks 17–24

e. weeks 25–36

BIRTH

_______ **13.** The mother's muscle contractions that help a baby to be born are called
   a. the uterus.
   b. the placenta.
   c. labor.
   d. full term.

_______ **14.** Your navel is the place on your body that was once attached to
   a. the uterus.
   b. the amnion.
   c. a fallopian tube.
   d. the umbilical cord.

FROM BIRTH TO DEATH

_______ **15.** Which of the following is a characteristic of infancy?
   a. Baby teeth appear.
   b. You begin to run.
   c. You begin to read.
   d. You need little sleep.

Directed Reading B *continued*

_______ **16.** Which of the following is a characteristic of childhood?
   a. You grow relatively slowly.
   b. You lose flexibility.
   c. You produce sex hormones.
   d. You grow permanent teeth.

17. You become an adolescent after you have reached _____________________.

18. What are two ways in which a male's body changes after puberty?

____

19. What are two ways in which a female's body changes after puberty?

____

20. What are two early signs of aging in adults under the age of 40?

____

21. What are two signs of aging in adults over the age of 65?

____

Vocabulary and Section Summary A

Human Reproduction

VOCABULARY

In your own words, write a definition of the following terms in the space provided.

1. testes

2. penis

3. ovary

4. uterus

5. vagina

SECTION SUMMARY

Read the following section summary.

- The male reproductive system produces sperm and can deliver sperm to the female reproductive system.
- The female reproductive system produces eggs, nurtures zygotes, and gives birth.
- If sperm are present in the female reproductive system within a few days of ovulation, fertilization may occur.
- A fertilized egg has one chromosome from each chromosome pair of the parents.
- Humans usually have one child per birth, but some people have multiple births.
- Human reproduction can be affected by infertility and by diseases such as cancer.

Skills Worksheet

Vocabulary and Section Summary A

Growth and Development

VOCABULARY

In your own words, write a definition of the following terms in the space provided.

1. embryo

____

____

2. placenta

____

____

3. pregnancy

____

____

4. umbilical cord

____

____

5. fetus

____

____

SECTION SUMMARY

Read the following section summary.

- Fertilization occurs when a sperm from the male joins with an egg from the female.

- First as an embryo and then as a fetus, a developing human undergoes many changes between implantation and birth.

- During the development of a human, cells differentiate.

- The umbilical cord and placenta support the developing human during pregnancy by providing oxygen and nutrients and by removing waste materials.

- The first stage of human development lasts from fertilization to birth.

- After birth, a human goes through four more stages of growth and development.

Vocabulary and Section Summary B

Human Reproduction

VOCABULARY

After you finish reading the section, try this puzzle! The underlined words below are scrambled. Write the completed words in the spaces provided.

1. The <u>SIPEN</u> transfers sperm from the male to the female.

   ____

2. The <u>YVOAR</u> produces eggs in the female.

   ____

3. Sperm are produced by the <u>STETSE</u>.

   ____

4. The <u>GAVANI</u> connects the outside of the body to the uterus.

   ____

5. A human embryo develops into a fetus in the <u>TURESU</u>.

   ____

Vocabulary and Section Summary B *continued*

SECTION SUMMARY

Read the following section summary.

- The male reproductive system produces sperm and can deliver sperm to the female reproductive system.

- The female reproductive system produces eggs, nurtures zygotes, and gives birth.

- If sperm are present in the female reproductive system within a few days of ovulation, fertilization may occur.

- A fertilized egg has one chromosome from each chromosome pair of the parents.

- Humans usually have one child per birth, but some people have multiple births.

- Human reproduction can be affected by infertility and by diseases such as cancer.

Vocabulary and Section Summary B

Growth and Development

VOCABULARY

After you finish reading the section, try this puzzle! Use the clues below to find the correct term. Then, find those words in the puzzle on the following page. Terms can be hidden in the puzzle vertically, horizontally, or backward.

__________________________ **1.** This is a partly fetal and partly maternal organ that exchanges materials between the mother and fetus.

__________________________ **2.** This begins on the first day of a woman's last menstrual period.

__________________________ **3.** This is a developing individual, from fertilization until the 10th week.

__________________________ **4.** This is a developing individual from the end of the 10th week until birth.

__________________________ **5.** This connects the fetus to the placenta.

▌Vocabulary and Section Summary B *continued*

S	Y	G	Q	E	V	Y	T	Z	A	H	Z	Y	S
F	D	P	L	V	A	D	K	S	T	Q	F	E	E
A	T	H	E	Z	J	K	J	E	Y	Z	E	H	A
J	H	A	T	Q	P	L	A	C	E	N	T	A	H
Z	Q	H	Y	Z	C	Z	Y	D	J	E	U	F	Y
Q	J	E	E	E	H	S	Q	X	Q	F	S	D	T
U	M	B	I	L	I	C	A	L	C	O	R	D	X
Y	V	A	C	A	H	T	H	D	J	Y	T	J	D
S	Y	E	Q	K	D	Z	T	E	H	R	D	Q	Y
G	D	X	S	J	A	V	S	V	Z	B	T	J	S
E	Q	T	K	Q	K	E	A	Y	Q	M	H	Q	H
A	D	H	E	J	D	T	H	G	V	E	V	A	T
Y	C	N	A	N	G	E	R	P	T	Y	J	S	Y
Q	D	C	S	F	Z	A	X	E	D	Z	Q	T	A

SECTION SUMMARY

Read the following section summary.

- Fertilization occurs when a sperm from the male joins with an egg from the female.
- First as an embryo and then as a fetus, a developing human undergoes many changes between implantation and birth.
- During the development of a human, cells differentiate.
- The umbilical cord and placenta support the developing human during pregnancy by providing oxygen and nutrients and by removing waste materials.
- The first stage of human development lasts from fertilization to birth.
- After birth, a human goes through four more stages of growth and development.

Reinforcement

Early Human Development

Complete this worksheet after you have finished reading the section "Growth and Development."

The following illustration shows the development of a human. Choose the term from the list below that best labels what is indicated in the diagram, and write the term in the corresponding box. Then, match each developing feature listed below to the stage at which it develops, and write the corresponding letter in the blank. Each feature and term are used once only. A stage may have more than one feature.

Terms

umbilical cord	fetus	birth
implantation	fertilization	placenta
amnion	embryo	

Developing Features

a. Tiny fingers and toes take shape.

b. Bones and bone marrow form.

c. Fetus responds to sunlight.

d. Heart begins to form.

e. Tiny muscle movements occur.

f. Blinking and swallowing occur.

g. Taste buds and eyebrows form.

h. Lungs "practice breathing."

i. Facial features look more human.

j. Brain and spinal cord begin to form.

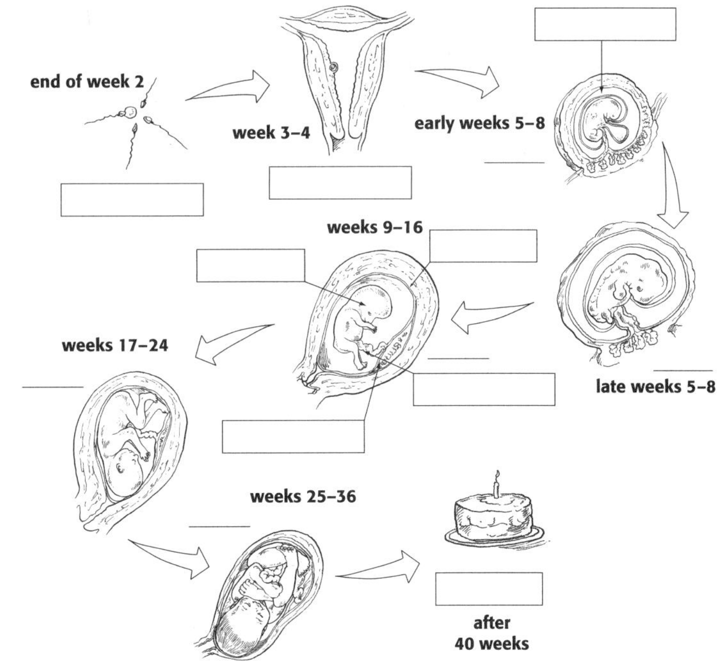

Critical Thinking

One to Grow On!

Aldous Huxley wrote *Brave New World* in 1932. This science fiction novel describes a society in the future. In this society, scientists manipulate the genes and the environment of embryos to determine a person's intelligence, physical strength, health, and social class.

In the novel, scientists add nutrients and oxygen to embryos that will become the ruling class and create oxygen deficiencies in embryos that will become members of the worker class. The people of this world are mass-produced in factories called hatcheries. The scientists produce thousands of individuals from one fertilized egg and sperm.

"For in nature it takes thirty years for two hundred eggs to reach maturity . . . we can make one ovary yield us over fifteen thousand adult individuals."

USEFUL TERMS

mass-produced describes something manufactured in large numbers

genetic engineering the manipulation of genes for practical purposes, such as identifying the genes that cause a particular disease

ANALYZING INFORMATION

1. How is the reproduction in *Brave New World* similar to the way twins form inside a human being? Explain your answer.

____

____

____

2. Would people "born" in the hatchery be considered fraternal or identical twins? Explain your answer.

____

____

____

____

Critical Thinking *continued*

UNDERSTANDING CONCEPTS

3. What functions normally performed by the placenta would the hatchery have to perform so that the embryos survive? Explain your answer.

____

____

____

4. Would modern reproductive disorders affect reproduction in the society described in *Brave New World?* Explain your answer.

____

____

____

____

AGREE OR DISAGREE

5. What are your thoughts about the genetic engineering described in *Brave New World?* Do you agree or disagree with the concept? Explain your answer.

____

____

____

____

____

____

____

____

SciLinks Activity

GROWTH AND DEVELOPMENT

Go to www.scilinks.org. To find links related to human development, type in the keyword HY70700. Then, use the links to complete the following activity about growth and development.

Internet Resources

For a variety of links related to this chapter, go to www.scilinks.org

Topic: Growth and Development
SciLinks code: HY70700

After you have checked out the SciLinks pages, imagine that you are either a child between the ages of two and five, or a retired older person, perhaps your own grandparent. Write a short story describing one particular day—it can be an ordinary day or some special day. In your story, write from the child's or the older adult's point of view. Use what you know about human growth and development to help you shape your story's content. Base your description on the SciLinks information. Next, in the table that follows, categorize the types of development included in your story as physical/language, emotional, or social.

____

____

____

____

____

____

____

____

____

____

____

Physical/Language	Emotional	Social

Section Review

Human Reproduction

USING VOCABULARY

1. Use *uterus* and *vagina* in the same sentence.

UNDERSTANDING CONCEPTS

2. **Identifying** What are the structures and functions of the male and female reproductive systems?

3. **Comparing** How is the production of sperm in the male reproductive system similar to the production of eggs in the female reproductive system?

4. **Describing** What are two reproductive system problems?

Section Review *continued*

CRITICAL THINKING

5. Predicting Consequences In females, an egg travels to the uterus through a fallopian tube approximately once a month. However, untreated STDs in women can block the fallopian tubes. How is fertilization affected in this situation?

____

____

____

____

____

6. Applying Concepts Twins can happen when a zygote splits in two or when two eggs are fertilized at the same time. Describe the difference between these situations in terms of how the offspring inherit genetic material from their father.

____

____

____

____

____

MATH SKILLS

7. Making Calculations In one country, 7 out of 1,000 infants die before their first birthday. Convert this statistic into a percentage. Show your work below.

Section Review *continued*

CHALLENGE

8. Identifying Relationships How do the male and female reproductive systems rely on other organ systems?

____

____

____

____

____

____

____

____

____

Section Review

Growth and Development

USING VOCABULARY

1. Write an original definition for *umbilical cord.*

2. Use *embryo* and *fetus* in the same sentence.

UNDERSTANDING CONCEPTS

3. Describing Outline the order of the development of tissues and organs in an embryo and fetus.

4. Summarizing Describe the processes of fertilization and implantation.

5. Listing What are five stages of human development?

Section Review *continued*

INTERPRETING GRAPHICS

Use the image below to answer the next question.

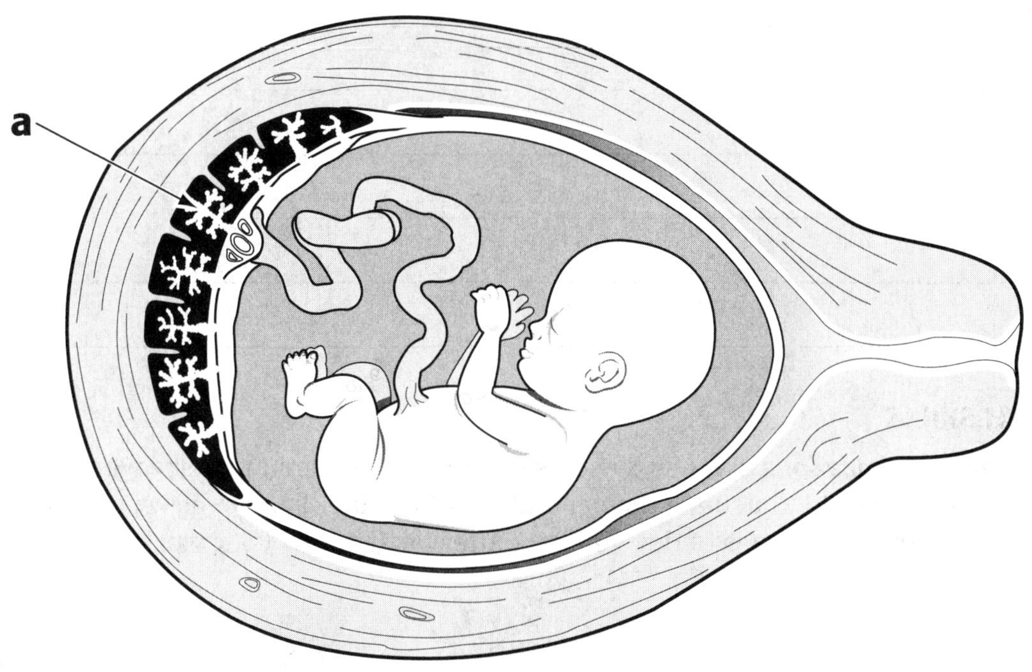

6. Evaluating What is the role of the structure indicated by **a**?

____

____

____

CRITICAL THINKING

7. Applying Concepts Why does the egg's covering change after a sperm has entered the egg?

____

____

____

____

____

____

____

Section Review *continued*

8. Analyzing Ideas Do you think that one stage of a human's life is more important than other stages are? Explain your answer.

____

____

____

____

____

MATH SKILLS

9. Making Calculations Alice is 80 years old, and she entered puberty when she was 12 years old. Calculate the percentage of her life that she has spent in each of the four stages of development after birth. Show your work below.

Skills Worksheet

Chapter Review

USING VOCABULARY

______ **1. Academic Vocabulary** In the sentence "Human reproductive organs in the female and male generate eggs and sperm," what does the word *generate* mean?
- **a.** make
- **b.** cause
- **c.** touch off
- **d.** spawn

For each pair of terms, explain how the meanings of the terms differ.

2. *embryo* and *fetus*

____

____

3. *testes* and *ovaries*

____

____

4. *uterus* and *vagina*

____

____

5. *fertilization* and *implantation*

____

____

6. *umbilical cord* and *placenta*

____

____

UNDERSTANDING CONCEPTS
Multiple Choice

______ **7.** Tissues and organs develop as an embryo becomes a fetus. Humans grow in size as they become adults. How are cells responsible for the growth that humans experience as they become adults?
- **a.** through cell division
- **b.** through cell expansion
- **c.** through cell death
- **d.** through cell contraction

Chapter Review *continued*

_______ **8.** All of the following are sexually transmitted diseases EXCEPT
    a. chlamydia.
    b. AIDS.
    c. infertility.
    d. genital herpes.

_______ **9.** The formation of identical twins occurs when
    a. a fertilized egg splits in two.
    b. two separate eggs are fertilized.
    c. implantation occurs.
    d. menstruation occurs.

_______ **10.** Which of the following is a function of the reproductive system?
    a. to produce all of the body's hormones
    b. to regulate body temperature
    c. to make hormones that fight disease
    d. to regulate the development of male and female characteristics

Short Answer

11. Identifying Which human reproductive organs produce sperm? Which produce eggs?

____

____

12. Describing Explain how the fetus gets oxygen and nutrients and how the fetus gets rid of wastes.

____

____

____

13. Summarizing What are four stages of human life after birth? Describe each stage.

____

____

____

____

14. Listing Name and describe three problems that can affect the human reproductive system.

____

____

____

Chapter Review *continued*

15. Modeling Draw a diagram showing the structures of the male and female reproductive systems. Label each structure, and explain how each structure contributes to fertilization and implantation.

16. Describing When do cells begin to differentiate in a developing human?

____

____

WRITING SKILLS

17. Writing from Research You have been asked to research the effects of vitamins on the growth and development of a human fetus. Develop a thesis for your research project. Then, briefly describe your project.

____

____

____

____

____

____

CRITICAL THINKING

18. Concept Mapping Use the following terms to create a concept map: *testes, penis, ovary, uterus, vagina, embryo, placenta, reproductive organs,* and *umbilical cord.*

19. Applying Concepts How do parents contribute genetic material to their offspring?

____

____

____

____

____

____

❙ Chapter Review *continued*

20. Making Inferences The birth of twins is the most common type of multiple birth—30 sets of twins are born for every 1,000 births. But the birth of quintuplets is very rare—1 set of quintuplets is born in about 53,000 births. Why might multiple births in which a large number of babies are born be less common than multiple births in which a small number of babies are born?

____

____

____

____

21. Drawing Conclusions Menstruation is affected by a hormone called *estrogen.* A woman who produces little estrogen may not have a menstrual cycle. In turn, the production of estrogen is affected by body fat. A woman who has little body fat usually produces less estrogen. What might happen to the menstrual cycle of a female athlete who exercises a lot?

____

____

____

____

INTERPRETING GRAPHICS

Use the image below to answer the next question.

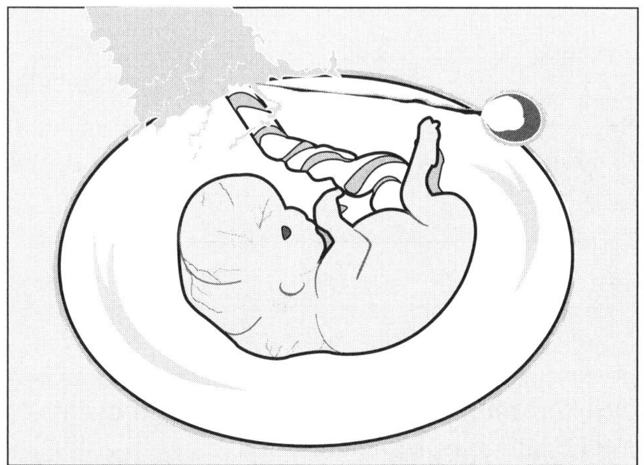

22. Identifying Relationships What is the name and function of the cord that connects the fetus to its mother?

____

____

INTERPRETING GRAPHICS

The following graph shows the cycles of the female hormone estrogen and the male hormone testosterone. The blue line shows the estrogen level in a female over the 28 days of her menstrual cycle. The red line shows the testosterone level in a male over the same time period.

Use the graph below to answer the next four questions.

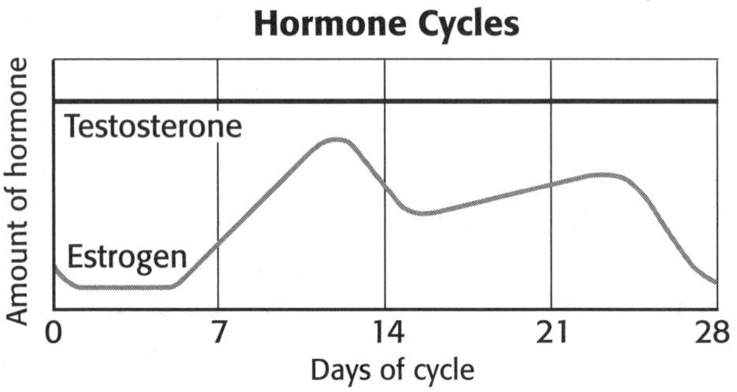

Hormone Cycles

23. Evaluating Data Over the 28 days, how do the day-to-day levels of testosterone differ from the day-to-day levels of estrogen?

____

____

____

24. Applying Concepts Compare the estrogen cycle to a woman's menstrual cycle. How are they related?

____

____

____

25. Making Inferences Why might the level of testosterone stay the same?

____

____

____

26. Making Inferences Do you think that the above estrogen cycle would change in a pregnant woman? Explain your answer.

____

____

____

| Chapter Review *continued*

MATH SKILLS

27. Making Calculations Identical twin births happen once in 100 births. How many sets of identical twins might you expect at a school that has 2,700 students? Show your work below.

CHALLENGE

28. Identifying Relationships How is the placenta specialized for its function?

____

____

____

____

____

____

____

____

____

____

____

____

____

____

Assessment

Chapter Pretest

Teacher Notes and Answer Key

The Pretest questions are designed to help you determine the prior knowledge of your students. Some questions test whether students have mastered the background knowledge they need to understand the content you are about to teach. Other questions test your students' prior knowledge of the content you are about to teach. Use the Pretest with the Test Doctors and diagnostic teaching tips in these notes pages to help you tailor your instruction to your students' specific needs.

QUESTION NUMBER	CORRECT ANSWER	STANDARD
1	D	7.5.a
2	C	7.5.a
3	D	7.5.g
4	A	7.5.g
5	B	7.5.a
6	D	7.1.f
7	B	7.2.b
8	B	7.5.d
9	C	7.5.d

TEST DOCTOR

The following Pretest questions have been diagnosed by the Test Doctor. Find out what might be causing your students' "ailing" answers. Each Test Doctor is followed by a diagnostic teaching tip to help you address students' learning needs.

Question 1 *asks students to identify the two parts of the nervous system that are responsible for maintaining homeostasis in the body.*

A **Incorrect.** The peripheral and central nervous systems encompass the entire nervous system. The peripheral nervous system refers to all the parts of the nervous system except the brain and the spinal cord, which make up the central nervous system.

B **Incorrect.** The brain and the spinal cord make up the central nervous system, which responds to all messages coming from the peripheral nervous system.

C **Incorrect.** The somatic and autonomic nervous systems are part of the peripheral nervous system. The somatic nervous system controls voluntary movements such as running and smiling, and the autonomic nervous system is responsible for involuntary body processes, such as digestion and heart rate.

D **Correct.** The sympathetic and parasympathetic nervous systems, part of the autonomic nervous system, work together to maintain homeostasis.

Diagnostic Teaching Tip: Students who have difficulty with this question should review the the role of the two divisions of the autonomic nervous system. Have students create a compare-and-contrast chart for the two divisions of the autonomic nervous system. Also review the meaning of the word *homeostasis*, and emphasize the role of the parasympathetic and the sympathetic nervous systems in maintaining homeostasis.

Chapter Pretest *continued*

Question 2 *asks students to identify the part of the brain that sends messages to the body to increase heart rate and blood pressure when they are frightened.*

A Incorrect. The cerebellum processes sensory information from the body so you can maintain balance and stand upright.

B Incorrect. The left hemisphere of the brain carries out conscious thoughts, actions, and learning. This hemisphere controls the right side of the body.

C Correct. The medulla sends messages through the autonomic nervous system to regulate and adjust body processes.

D Incorrect. The right hemisphere is part of the cerebrum and controls spatial thinking and the activities consciously conducted on the left side of the body.

Diagnostic Teaching Tip: Students who answer this question incorrectly should review the roles of each part of the brain with a partner. Have them identify the parts responsible for conscious behavior and learning, body position, and involuntary body processes. Have pairs focus on the interaction between the medulla and autonomic nervous system. Students should learn that the medulla sends signals through the autonomic nervous system to influence the speed of body functions.

Question 3 *asks students to identify the part of the eye that lets light in.*

A Incorrect. The cornea is a membrane that covers and protects the eye.

B Incorrect. The retina is a layer of light-sensitive cells called photoreceptors that are located at the back of the eye.

C Incorrect. The lens focuses images on the retina.

D Correct. The pupil is the opening that lets light in the eye.

Diagnostic Teaching Tip: Students who have trouble answering this question correctly should review visuals of the components of the eye and list the major function of each part. Have students review the path light from a light source takes as it travels through the eye's tissues and study the part each tissue plays in the process of sensing and comprehending a visual image.

Question 4 *asks students to identify the part of the ear that stimulates auditory neurons to send sound impulses to the brain.*

A Correct. The cochlea is filled with fluid that vibrates and stimulates the auditory nerve. These impulses are sent to the brain and interpreted as sound.

B Incorrect. The eardrum in the middle ear makes three small bones in the middle ear vibrate.

C Incorrect. The hammer is one of the ear bones that are made to vibrate by the eardrum.

D Incorrect. The stirrup is the bone in the middle ear that vibrates against the cochlea to create liquid vibrations inside the cochlea.

Chapter Pretest *continued*

Diagnostic Teaching Tip: Students who have trouble with this question should study the different components of the human ear. Have students list the specific function of each part and how it interacts with the other components to control hearing and balance. Students should indicate which components are part of the outer ear, the middle ear, and the inner ear.

Question 5 *asks students to explain how skin sensory nerves can distinguish between different types of stimuli.*

A Incorrect. Each receptor responds to one type of stimulus.
B Correct. The stimulation that is experienced depends upon the specific receptor that is being stimulated.
C Incorrect. The type of sensation experience does not depend upon the length of time a receptor is stimulated.
D Incorrect. The specific type of sensation experienced does not depend on how hard the receptor is affected by the stimulus.

Diagnostic Teaching Tip: Students who have trouble answering this question should discuss the types of sensory receptors in the skin. Discussion should focus on the different shape and placement of receptors in the integumentary tissue and the resulting sensations that are sent to the brain for processing.

Question 6 *asks students to demonstrate an understanding of the progression of the development of a baby from the point of conception through birth.*

A Incorrect. The zygote is the first stage after the fertilization of the egg.
B Incorrect. The embryo stage occurs after the zygote stage.
C Incorrect. The embryo stage occurs before the fetus stage.
D Correct. The egg develops into a zygote, then an embryo, a fetus, and finally an infant at birth.

Diagnostic Teaching Tip: Students who have difficulty with this question should be introduced to the timeline of development for an unborn human baby. Focus on the major time periods between fertilization, development into an embryo, and maturity into a fetus. Discuss important tissue and organ developments that occur. Be sure to emphasize Section 2, "Growth and Development," in Chapter 18, "Reproduction and Development." Students must know that as multicellular organisms like an unborn human baby develop, their cells differentiate, in order to master standard 7.1.f.

Question 7 *asks students to know how many chromosomes are in a male sperm cell compared to how many chromosomes are in a skin cell.*

A Incorrect. Both the male and female sex cells contribute one half the total number of chromosomes that are present in the zygote or in other body cells, such as a skin cell.
B Correct. Since both the male and female parents contribute chromosomes during fertilization, the number contributed by each is one half of the full number present in a body cell, such as a skin cell.

Chapter Pretest *continued*

C Incorrect. Sperm does not contain twice the number of chromosomes possessed by a skin cell.

D Incorrect. Chromosomes are developed and replicated during fertilization.

Diagnostic Teaching Tip: Students who have difficulty answering this question correctly should review the process of meiosis and the second cell division. Emphasize the fact that during fertilization two different parents contribute chromosomes to create an organism that has the same number of chromosomes as each parent. Be sure to emphasize Section 1, "Human Reproduction," in Chapter 18, "Reproduction and Development." Students must know that sexual reproduction produces offspring that inherit half their genes from each parent in order to master standard 7.2.b.

Question 8 *asks students to identify the first thing that occurs after sperm successfully penetrates an egg during the process of fertilization.*

A Incorrect. After fertilization, the egg develops a boundary to keep other sperm from entering.

B Correct. After fertilization, the egg's coating changes to prevent penetration by other sperm.

C Incorrect. The egg only splits into two identical cells with twins. This happens after the sperm and egg nuclei fuse.

D Incorrect. After fertilization the sperm and egg cell fuse into one and continue to go through cell division.

Diagnostic Teaching Tip: If students have difficulty with this question, have the class create a timeline starting prior to fertilization through fertilization and the implantation of the zygote in the uterine wall. Use visuals to show what happens at each stage from the development and release of the egg through fertilization. Be sure to emphasize Section 2, "Growth and Development," in Chapter 18, "Reproduction and Development." Students must be able to understand the process of fertilization in human sexual reproduction in order to master standard 7.5.d.

Question 9 *asks students to identify in which part of the female reproductive system fertilization of the egg takes place.*

A Incorrect. The ovary is the organ that produces eggs.

B Incorrect. The uterus is the organ where the embryo develops into a fetus.

C Correct. The fallopian tube is the organ in which sperm fertilizes an egg.

D Incorrect. The vagina is the organ that connects the outside of the body to the uterus.

Diagnostic Teaching Tip: Students who have difficulty with this question should be introduced to a visual that follows the formation of an egg to its fertilization through its implantation in the wall of the uterus. Be sure to emphasize Section 2, "Growth and Development," in Chapter 18, "Reproduction and Development." Students must be able to understand fertilization in order to master standard 7.5.d.

Chapter Pretest

______ **1.** Two parts of the nervous system are responsible for working together
to maintain homeostasis, a stable internal state in the body. What are
the two divisions of the nervous system?
A peripheral and central nervous systems
B brain and the spinal cord
C somatic and autonomic nervous systems
D sympathetic and parasympathetic nervous systems

______ **2.** When you are frightened by something, what part of the brain directs
your body to react by making your heart beat faster and increasing
your blood pressure?
A cerebellum
B left hemisphere
C medulla
D right hemisphere

______ **3.** What part of the human eye allows light to enter the eye?
A cornea
B retina
C lens
D pupil

______ **4.** What part of the human ear is responsible for stimulating the neurons
that relay the sensation of sound to the brain?
A cochlea
B eardrum
C hammer
D stirrup

______ **5.** How can your body tell the difference between a tickle and a
pin prick?
A from the stimulation of more than one receptor
B from what kind of receptor is being stimulated
C from how long the receptor is being stimulated
D from how hard the receptor is being stimulated

______ **6.** An infant's development from fertilization of the egg to birth involves a
series of stages. These stages occur in a specific order. Which
of the following shows the order from release of an egg to the birth
of an infant?
A egg, fetus, zygote, embryo, infant
B egg, embryo, fetus, zygote, infant
C egg, zygote, fetus, embryo, infant
D egg, zygote, embryo, fetus, infant

Chapter Pretest *continued*

_______ **7.** How many chromosomes are present in a sperm cell in comparison to
a skin cell?
   A the same number as the skin cell has
   B half as many chromosomes as the skin cell has
   C twice as many chromosomes as the skin cell has
   D less because the sperm cell's chromosomes are not yet developed

_______ **8.** What is the first thing that happens when a sperm has successfully
penetrated an egg?
   A The coating of the egg allows other sperm to enter.
   B The coating of the egg changes to prevent other sperm from
   entering.
   C The egg splits and forms two identical egg cells.
   D The sperm and egg cells no longer go through cell division.

_______ **9.** At which part of the following diagram of the female reproductive
system does fertilization of the egg occur?

   A ovary
   B uterus
   C fallopian tube
   D vagina

Name _______________________________ Class _______________ Date _______________

Section Quiz

Section: Human Reproduction

Write the letter of the correct answer in the space provided.

______ **1.** Sperm, the male sex cells, are produced in which organ?
    a. testes   **c.** prostate
    b. epididymis   **d.** bladder

______ **2.** Semen leaves the male body through which organ?
    a. the penis   **c.** the urethra
    b. the vas deferens   **d.** the bladder

______ **3.** A female's eggs are produced in which organ?
    a. the uterus   **c.** the ovary
    b. the vagina   **d.** the fallopian tube

______ **4.** Where, in mammals, does the zygote develop?
    a. in the vagina   **c.** in the urethra
    b. in the uterus   **d.** in the fallopian tube

______ **5.** Which one of the following is the canal through which a baby passes when it is born?
    a. the uterus   **c.** the fallopian tube
    b. the ovary   **d.** the vagina

______ **6.** What is the function of the menstrual cycle?
    a. to destroy unfertilized eggs
    b. to prevent ovulation
    c. to prepare the body for pregnancy
    d. to ready the vagina for fertilization

______ **7.** In fertilization, how many chromosomes come from each parent?
    a. one   **c.** three
    b. two   **d.** four

______ **8.** What is an STD?
    a. a hormone
    b. an influenza virus
    c. a disease transmitted by sex
    d. a disease transmitted through pregnancy

______ **9.** Sexual activity that releases only a few sperm
    a. can lead to fertilization but not pregnancy.
    b. can lead to fertilization and pregnancy.
    c. cannot lead to fertilization.
    d. cannot lead to pregnancy.

Section Quiz

Section: Growth and Development

Write the letter of the correct answer in the space provided.

_______ **1.** When a fertilized egg divides and forms a ball of cells, it is known as
- **a.** a nucleus.
- **b.** an implant.
- **c.** an embryo.
- **d.** a fetus.

_______ **2.** Where does implantation occur and the zygote develop?
- **a.** in the ovary
- **b.** in the uterus
- **c.** in the vagina
- **d.** in the fallopian tube

_______ **3.** The developing embryo is fed by its mother's blood through the
- **a.** placenta.
- **b.** amnion.
- **c.** vagina.
- **d.** uterus.

_______ **4.** What cells begin to form early on in the developing embryo?
- **a.** lung cells
- **b.** ear cells
- **c.** eye cells
- **d.** blood cells

_______ **5.** The developing embryo is connected to the placenta by
- **a.** the spinal cord.
- **b.** the amniotic fluid.
- **c.** the umbilical cord.
- **d.** the developing nerves.

_______ **6.** After week 9 or 10 of pregnancy, the developing human acquires features and is then referred to as
- **a.** a zygote.
- **b.** a fetus.
- **c.** a newborn.
- **d.** an embryo.

_______ **7.** A baby is usually ready to be born after how many weeks?
- **a.** 17
- **b.** 25
- **c.** 40
- **d.** 48

_______ **8.** A child is considered an infant until it reaches what age?
- **a.** 1 year
- **b.** 2 years
- **c.** 3 years
- **d.** 5 years

_______ **9.** During puberty, an adolescent's body goes through what change?
- **a.** Permanent teeth appear.
- **b.** The reproductive system matures.
- **c.** There is loss of flexibility.
- **d.** Eyesight deteriorates.

_______ **10.** At what pregnancy stage do the embryo cells begin to differentiate?
- **a.** weeks 1 and 2
- **b.** weeks 3 and 4
- **c.** weeks 5 to 8
- **d.** weeks 9 to 16

Chapter Test A

Reproduction and Development
MULTIPLE CHOICE
Write the letter of the correct answer in the space provided.

______ **1.** Semen exits the male body through which organ?
    a. the testis
    b. the vas deferens
    c. the epididymis
    d. the penis

______ **2.** The sperm and egg each contribute how many chromosomes to the fertilized egg?
    a. one
    b. two
    c. three
    d. four

______ **3.** Which organ connects the outside of the female body to the uterus?
    a. the ovaries
    b. the oviducts
    c. the vagina
    d. the fallopian tubes

______ **4.** After week five, the embryo receives nutrients from the placenta through which of the following?
    a. the vagina
    b. the umbilical cord
    c. the fallopian tubes
    d. the urethra

______ **5.** When is a woman's pregnancy commonly considered to start?
    a. the day before the last menstrual period
    b. the first day of the last menstrual period
    c. two weeks after the last menstrual period
    d. on the day fertilization takes place

______ **6.** Which male organ produces sperm?
    a. the penis
    b. the urethra
    c. the ovary
    d. the testis

▌Chapter Test A *continued*

_______ **7.** Which female organ produces eggs?
- **a.** the vagina
- **b.** the uterus
- **c.** the bladder
- **d.** the ovary

_______ **8.** A fertilized egg develops in which female organ?
- **a.** the uterus
- **b.** the fallopian tube
- **c.** the vagina
- **d.** the ovulation

_______ **9.** What is the 28-day period during which a female's body prepares for pregnancy?
- **a.** puberty
- **b.** labor
- **c.** ovulation cycle
- **d.** menstrual cycle

_______ **10.** Two babies born together that look exactly alike are called what?
- **a.** fraternal twins
- **b.** identical twins
- **c.** quintuplets
- **d.** quadruplets

_______ **11.** What is one sexually transmitted disease a person can get during sexual activity?
- **a.** the flu
- **b.** zygote
- **c.** pneumonia
- **d.** chlamydia

_______ **12.** When cells grow in an uncontrolled way, what may develop?
- **a.** an embryo
- **b.** a cancer
- **c.** a zygote
- **d.** hepatitis

_______ **13.** Couples who cannot have children are said to be what?
- **a.** infertile
- **b.** in labor
- **c.** asexual
- **d.** ovulating

_______ **14.** Differentiation of an embryo's cells begins during which of the
following weeks of pregnancy?
  a. 1 and 2
  b. 3 and 4
  c. 5 through 8
  d. 9 through 16

_______ **15.** Which of the following statements is true about sexual activity without
ejaculation?
  a. It cannot release sperm.
  b. It cannot lead to fertilization.
  c. It can release a few sperm and lead to pregnancy.
  d. It can release a few sperm but cannot lead to pregnancy.

MATCHING

**Match the correct description with the correct term. Write the letter in the space
provided.**

_______ **16.** the fertilized egg after it is implanted in the
uterus

_______ **17.** a developing human after 9 or 10 weeks of
pregnancy

_______ **18.** connects the fetus to the placenta

a. umbilical cord

b. fetus

c. embryo

**Match the correct description with the correct term. Write the letter in the space
provided.**

_______ **19.** This grows quickly during weeks 7 to 8 of
pregnancy.

_______ **20.** A 19-week-old fetus can respond to this.

_______ **21.** A mother has contractions as a baby is born.

a. brain

b. labor

c. sound

FILL-IN-THE-BLANK

Use the terms from the following list to complete the sentences below.

infant	childhood
adolescence	young adulthood

22. During _______________________, a person's reproductive system matures.

23. A child is called a(n) _______________________ until it is 2 years old.

24. When people are in _______________________, they are at the peak of their physical development.

25. During _______________________, you grow a set of permanent teeth.

Assessment

Chapter Test B

Reproduction and Development
MULTIPLE CHOICE
Write the letter of the correct answer in the space provided.

_______ **1.** The male hormone, testosterone, is produced in the
   a. bladder.
   b. urethra.
   c. testes.
   d. penis.

_______ **2.** A male's sperm mixes with fluid that is made in the
   a. penis.
   b. prostate gland.
   c. vas deferens.
   d. scrotum.

_______ **3.** Semen leaves the body through the penis via the
   a. urethra.
   b. epididymis.
   c. bladder.
   d. testis.

_______ **4.** One sexually transmitted disease that a person can get during sex with an infected person is
   a. cancer.
   b. liver disease.
   c. influenza.
   d. herpes.

_______ **5.** A man who does not produce enough healthy sperm usually
   a. has hepatitis.
   b. is infertile.
   c. has cancer.
   d. is an identical twin.

_______ **6.** A female's eggs are produced in the
   a. uterus.
   b. ovaries.
   c. fallopian tube.
   d. vagina.

_______ **7.** A female's egg is fertilized when it is in the
   a. fallopian tube.   **c.** ovary.
   b. bladder.   **d.** urethra.

Chapter Test B *continued*

_______ **8.** Implantation takes place when the embryo embeds itself in the wall of
the
   a. vagina.
   b. ovary.
   c. fallopian tube.
   d. uterus.

_______ **9.** When a baby is born, it passes out of the mother's body through the
   a. urethra.
   b. fallopian tube.
   c. vagina.
   d. uterus.

_______ **10.** If her egg is not fertilized, a woman's uterus sheds blood and tissue on
the first day of
   a. ovulation.
   b. the menstrual cycle.
   c. fertilization.
   d. the hormonal cycle.

_______ **11.** In a multiple birth, if two offspring who look alike and share the same
genes are born, they are called
   a. fraternal twins.
   b. identical twins.
   c. multiples.
   d. triplets.

_______ **12.** One of the rarest type of multiple births, occurring once in every
53,000 births, produces five babies, who are called
   a. identical twins.
   b. triplets.
   c. quintuplets.
   d. quadruplets.

_______ **13.** Two of the most common reproductive cancers among women are
cancer of the cervix and
   a. breast cancer.
   b. cancer of the ovary.
   c. cancer of the uterus.
   d. fallopian cancer.

_______ **14.** Differentiation of an embryo's cells begins during weeks
   a. 1 and 2 of pregnancy.
   b. 3 and 4 of pregnancy.
   c. 5 through 8 of pregnancy.
   d. 9 through 16 of pregnancy.

_______ **15.** Sexual activity without ejaculation
    a. cannot release sperm.
    b. cannot lead to fertilization.
    c. can release a few sperm and lead to pregnancy.
    d. can release a few sperm but cannot lead to pregnancy.

MATCHING

Match the correct description with the correct term. Write the letter in the space provided.

_______ **16.** quick growth, baby teeth appear, you begin to walk

_______ **17.** peak of physical development

_______ **18.** reproductive system matures, body changes into adult form

_______ **19.** permanent teeth grow in, greater muscle coordination and activity

_______ **20.** gray hair appears, physical abilities begin to decline

_______ **21.** wrinkles may become prominent, physical strength continues to decline

    a. childhood
    b. young adult
    c. older adult
    d. infancy
    e. adolescence
    f. middle age

Match the correct description with the correct term. Write the letter in the space provided.

_______ **22.** process in which each parent provides one chromosome to the egg

_______ **23.** organ that connects the placenta to the fetus

_______ **24.** what the embryo develops into after the 10th week

_______ **25.** a fertilized egg

_______ **26.** organ that exchanges materials between the mother and the fetus

_______ **27.** period of time between the first day of the last menstrual period and birth

    a. embryo
    b. placenta
    c. pregnancy
    d. umbilical cord
    e. fertilization
    f. fetus

Match the labels to the drawing. Write the letters in the spaces provided.

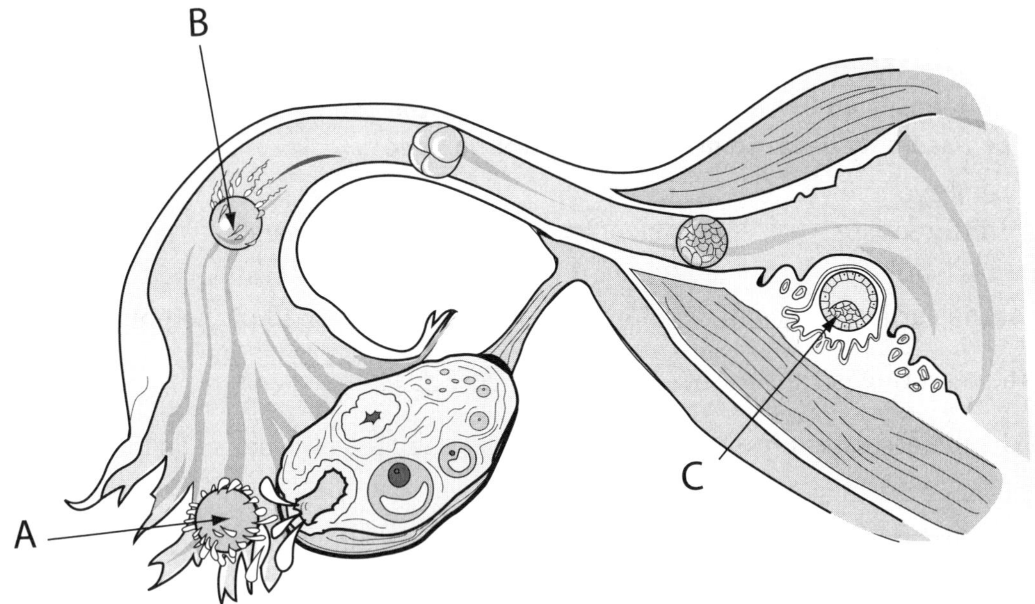

_______ **28.** The embryo implants itself in the wall of the uterus.

_______ **29.** The egg is released from the ovary.

_______ **30.** The egg is fertilized in the fallopian tube by the sperm.

Chapter Test C

Reproduction and Development

USING KEY TERMS

Use the terms from the following list to complete the sentences below. Each term may be used only once. Some terms may not be used.

embryo	sperm	ovary
differentiate	placenta	pregnancy
uterus	fetus	fertilization

1. At the earliest stages of development, before week 9 or 10 of the pregnancy, the offspring is called a(n) ____________________.

2. The male's testes produce ____________________, the sex cells that carry his genes.

3. The first day of the last menstrual period is the beginning of ____________________, which usually ends around the 40th week.

4. In mammals, the offspring develops after the zygote is implanted in the ____________________, whose walls become thick and rich in nutrients.

5. The developing offspring obtains nutrients via the ____________________, which is an organ of two-way exchange between the mother and the offspring.

6. The developing offspring is called a(n) ____________________ after week 9 or 10 of pregnancy.

7. During weeks 3 and 4, an embryo's cells begin to specialize or ____________________.

UNDERSTANDING KEY IDEAS

Write the letter of the correct answer in the space provided.

______ **8.** The organ that transfers sperm from the male to the female is the
  a. epididymis.
  b. ovary.
  c. penis.
  d. testis.

Chapter Test C *continued*

______ **9.** Ovulation occurs on about the 14th day of the
  a. ovarian cycle.
  b. menstrual cycle.
  c. formation of the placenta.
  d. sperm fertilization.

______ **10.** The genetic information passed on to offspring is carried by the sperm's and egg's
  a. chromosomes.
  b. fetus.
  c. meiosis.
  d. amnion.

______ **11.** The embryo is connected to the placenta via the
  a. fallopian tube.
  b. zygote.
  c. monotreme.
  d. umbilical cord.

______ **12.** The testes produce
  a. estrogen.
  b. progesterone.
  c. testosterone.
  d. semen.

______ **13.** A person's reproductive system matures during
  a. childhood.
  b. young adulthood.
  c. implantation.
  d. puberty.

14. What hormones do the ovaries produce? What is the function of these hormones?

____

____

15. Where are sperm made, and how do they leave the body?

____

____

16. What are the differences between fraternal twins and identical twins?

____

____

____

17. What are the differences between an embryo and a fetus?

____

____

18. Can sexual activity without ejaculation lead to pregnancy? Explain your answer.

____

____

CRITICAL THINKING

19. Why is it crucial for sexually reproducing organisms to produce sex cells that have half the normal number of chromosomes?

____

____

____

____

____

20. A newborn's skull is made up of bone plates connected by soft, flexible material that allows the skull to change shape. Why is this characteristic important?

____

____

____

____

Chapter Test C *continued*

CONCEPT MAPPING

21. Use the following terms to complete the concept map below:

vas deferens	fallopian tubes	epididymis
uterus	testes	ovaries

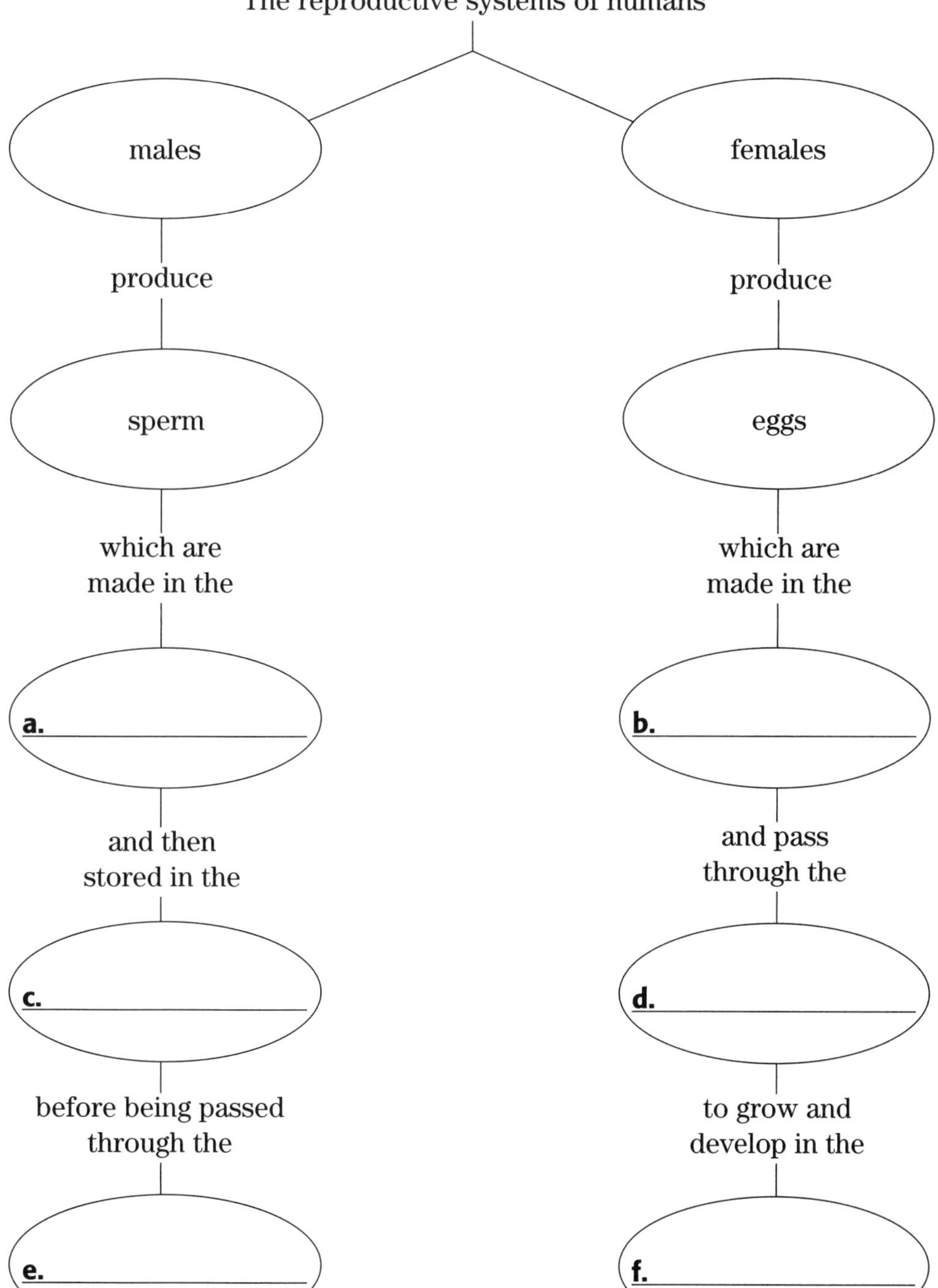

Performance-Based Assessment

Teacher Notes

PURPOSE

Students will make posters to illustrate the major stages of human development and the events associated with each stage.

Gladys Cherniak
St. Paul's Episcopal School
Mobile, Alabama

TIME REQUIRED

One 45-minute class period

RATING

Easy ← 1 2 3 4 → Hard

Teacher Prep–1
Student Set-Up–1
Concept Level–2
Clean Up–1

ADVANCE PREPARATION

Ahead of time, gather a collection of magazines that have many pictures of people at all stages of life, including some that show the developing embryo and fetus, if possible. In case the magazines do not contain all the necessary pictures of the different stages of human development, provide copies of appropriate pictures from the textbook. Have enough different-colored markers or pencils for all students.

SAFETY INFORMATION

Students should handle scissors with care.

TEACHING STRATEGIES

This activity works best in groups of 2–3 students.

Performance-Based Assessment *continued*

Evaluation Strategies

Use the following rubric to help evaluate student performance.

Rubric for Assessment

Possible points	Quality of poster and clarity of descriptions (80 points possible)
80–60	Excellent display; clear and accurate labeling of stages; high level of detail
59–50	Good display; clear and accurate labeling of structures; good level of detail
49-20	Fair display; inaccurate labeling of stages; some attention to detail
19–1	Erroneous, incomplete, or unclear descriptions
	Completion of analysis (20 points possible)
20–10	Superior understanding of human development; uses correct terminology; answers expressed clearly and accurately
9–1	Fair understanding of human development; uses correct terminology; answers unclear and inaccurate

Assessment

Performance-Based Assessment

OBJECTIVE

You will illustrate the major stages of human development and the events associated with each stage.

KNOW THE SCORE!

As you work through the activity, keep in mind that you will be earning a grade for the following:

- the quality of your poster and the clarity of your descriptions (80%)
- how well you complete the analysis questions (20%)

MATERIALS

- magazines
- markers or pencils, colored
- paste or tape
- poster board or white paper
- scissors

SAFETY INFORMATION 

- Be careful when using scissors.

PROCEDURE

1. Find examples in the magazines of the different stages of human development from fertilization through older adulthood, and cut out one or two pictures to represent each stage. Draw pictures for any stage that is not represented by a magazine image.

2. Arrange the pictures on your paper or poster board. Glue or tape them in place.

3. Underneath each picture, write the name of the stage it represents, along with a description of the major events that happen during that stage.

ANALYZE THE RESULTS

4. Which developmental stage is generally the longest?

____

5. Why was it difficult for you to tie your shoes before age 4 or 5?

____

____

Performance-Based Assessment *continued*

6. How would your life be affected if your development suddenly stopped now?

____

____

7. Why is proper development prior to puberty important, and how does it affect your life as an adult?

____

____

____

BIG IDEA QUESTION

8. At what stage of life can humans typically begin to reproduce? Explain your answer.

____

____

____

Assessment

Standards Assessment

Teacher Notes and Answer Key

To provide practice under more realistic testing conditions, give students 20 min to answer all of the questions in this assessment.

QUESTION NUMBER	CORRECT ANSWER	STANDARD
1	C	7.5.e (supporting)
2	A	7.5.a (supporting)
3	B	7.5.d (supporting)
4	D	7.2.b (supporting)
5	D	7.1.f (supporting)
6	A	7.5.a (mastering)
7	A	7.5.d (mastering)
8	B	7.5.d (mastering)
9	D	7.5.d (exceeding)
10	A	7.2.b (mastering)
11	A	7.5.a (mastering)
12	A	7.1.c (mastering)
13	A	7.2.e (mastering)
14	C	7.2.c (mastering)
15	B	7.1.c (mastering)

TEST DOCTOR

The following Standards Assessment questions have been diagnosed by the Test Doctor. Find out what might be causing your students' "ailing" answers. Each Test Doctor is followed by a diagnostic teaching tip to help you address students' learning needs.

Question 1 *asks students to define the word* function.

A Incorrect. A *position* is a location or arrangement.

B Incorrect. The placenta is an *organ*, but that is not the meaning of the word *function*.

C Correct. The placenta serves a certain *purpose* when a baby is developing. It provides food and oxygen for the baby.

D Incorrect. A *result* is the consequence of an action.

Diagnostic Teaching Tip: Students who answer this question incorrectly might need to review the functions of different organs. Give students a list of different organs such as heart, lungs, stomach, brain, ovary. Then, have them list the function that each organ performs.

Question 2 *asks students to choose the correct word to complete a sentence about the uterus.*

A Correct. The uterus is a special structure that is part of the female reproductive system. It is the organ in which a baby develops.

B Incorrect. The uterus is not a *system;* it is part of the reproductive system.

Standards Assessment *continued*

C **Incorrect.** The uterus is not an *organism*. It can't sustain life on its own.

D **Incorrect.** The uterus is not a way of doing something, or a *method*.

Diagnostic Teaching Tip: Students who answer this question incorrectly might need to review the structure of organs and tissues within the female reproductive system, including the uterus. Encourage students to explore the relationship of structure and function by asking them to explain why the uterus is a good "house" for the baby as it develops.

Question 3 *asks students to define the word* generate.

A **Incorrect.** To *generate* is to bring about or to produce; to *explain* is to describe or to make something understandable.

B **Correct.** To *generate* something means to *produce* it. The bodies of males and females generate sperm and eggs, respectively.

C **Incorrect.** To *generate* is to bring about or to produce; to *identify* is to locate or select something based on specific characteristics.

D **Incorrect.** To *generate* does not mean to *accomplish*. To *accomplish* is to finish or succeed in doing something.

Diagnostic Teaching Tip: Students who answer this question incorrectly might benefit from a review of how the male body generates sperm. Have students list the steps required for a sperm to be generated.

Question 4 *asks students to choose the correct group of words that best completes a sentence.*

A **Incorrect.** *Asexual reproduction* is a type of reproduction that involves only one parent organism. The offspring will be identical to this parent. Methods of asexual reproduction include budding, mitosis, and propagation from runners or plantlets.

B **Incorrect.** *Sexual function* refers to how an organism functions sexually, and is not the best choice within this context.

C **Incorrect.** *Sexual selection* refers to evolution and is not appropriate within this context.

D **Correct.** *Sexual reproduction* is the best group of words to use in the sentence. It is through sexual reproduction that offspring inherit genes from two parents. The traits of the offspring of sexual reproduction are a combination of the traits of both parents.

Diagnostic Teaching Tip: Students who answer this question incorrectly may need to review sexual reproduction. Have students compare asexual reproduction and sexual reproduction.

Question 5 *asks students to choose a word that matches a scientific definition.*

A **Incorrect.** Production refers to the making or manufacture of cells, but not specialization.

B **Incorrect.** Fabrication does mean to produce or build, but it does not describe the specialization of cells.

C Incorrect. Cells do show distinction from each other through specialization, but this does not describe the process by which they do it.

D Correct. Differentiation is the process by which parts of an organism changes to enable specialization of those parts.

Diagnostic Teaching Tip: Students who answer this question incorrectly might need to review cell differentiation. Have students list a body system. Then, ask students to work backwards to list all the organs, tissues, and cells within the system. Point out that each cell has a certain function to perform that no other cell can perform.

Question 6 *asks students to demonstrate an understanding of functions during pregnancy.*

A Correct. The placenta is connected to the fetus through an umbilical cord. It is a two-way exchange organ. It provides nutrients and oxygen to the fetus and removes waste.

B Incorrect. The uterus is the organ that holds the fetus.

C Incorrect. The thin membrane is called the amnion.

D Incorrect. The placenta allows for nutrients from the mother to pass to the fetus, and for wastes from the fetus, not nutrients, to pass to the mother.

Diagnostic Teaching Tip: Students who struggle with this question might benefit from reviewing the different structures found within the uterus during pregnancy. Give students a picture of a fetus in the womb, and have them label and describe the different parts.

Question 7 *asks students to demonstrate understanding of the cause of pregnancy.*

A Correct. Sperm move through the vagina, uterus, and a fallopian tube. If an egg is present, a sperm may pierce the egg and fertilize it. This fertilization event is the beginning of pregnancy

B Incorrect. A menstrual cycle is the body's way of preparing for a possible pregnancy. A menstrual cycle does not cause pregnancy.

C Incorrect. A man's ejaculate may contain sperm, even during puberty. For a pregnancy to result, a sperm must fertilize an egg. Therefore, in order for ejaculation to cause pregnancy, ejaculation must occur in or near a woman's vagina.

D Incorrect. Ovulation is the release of an egg from an ovary. Pregnancy will not occur unless the egg is fertilized by a sperm. Ovulation alone will not cause pregnancy.

Diagnostic Teaching Tip: Students who struggle with this question might benefit from reviewing fertilization and the actions preceding fertilization. Begin with diagrams of the male and female reproductive systems. Ask students to identify where sperm and eggs are held and the paths they travel. Then, have students describe the process of fertilization using scientific terminology and referencing the diagrams.

Standards Assessment *continued*

Question 8 *asks students to know the male reproductive system.*

A Incorrect. This labels the skin sac, which holds the testes.
B Correct. These are the testes. They are the pair of organs that make sperm and testosterone.
C Incorrect. This is the vas deferens tube, where fluids mix with the sperm.
D Incorrect. This is the penis, the organ through which sperm exits a man's body.

Diagnostic Teaching Tip: Students who struggle with this type of question might benefit from reviewing the male reproductive system. Give students a picture of the male reproductive system, and have them label and describe the functions of the different parts.

Question 9 *asks students to demonstrate an understanding of fertilization.*

A Incorrect. The sperm does have to penetrate the outer coating but at this point the egg isn't considered fertilized.
B Incorrect. In order for fertilization to occur, an egg and sperm need to be present, but it doesn't ensure fertilization.
C Incorrect. A few sperm do usually cover the outer coating but this does not mean fertilization has occurred.
D Correct. Fertilization occurs when the nucleus of the egg joins with the nucleus of the sperm. The cell now has both sets of chromosomes and can start to divide.

Diagnostic Teaching Tip: Students who answer this question incorrectly might benefit from a review of fertilization. Have students describe or make a flowchart of the process of fertilization.

Question 10 *asks students to understand that offspring inherit half their genes from each parent.*

A Correct. An egg and sperm both contain half of the genetic material needed to create a cell. When an egg is fertilized, the DNA from the sperm joins with the DNA of the egg to make one cell. So half of the baby's genetic material is from the father and half is from the mother.
B Incorrect. The genes are not mixed up in the uterus; instead, the genes are combined at fertilization when the nuclei of the egg and sperm combine. The uterus is the site of growth for the fetus.
C Incorrect. The very first cell of the new organism, the zygote, contains the genetic material of both parents. The zygote splits into two cells, and these cells split again and so on. Cells do differentiate during development, but this does not affect the genetic material in the cells.
D Incorrect. The environment doesn't typically determine physical characteristics of a child.

Diagnostic Teaching Tip: Students who answered this question incorrectly might need to review how offspring inherit genetic traits from their parents. Have students list specific traits that they have such as eye color, hair color, height, shape of nose, or curling of tongue. Then, have them decide whether or not they received the trait from their father's side, mother's side, or a combination.

Standards Assessment *continued*

Question 11 *asks students to know the female reproductive system.*

A Correct. Ovary best fits into the space because ovaries make the eggs and produce the hormones estrogen and progesterone.

B Incorrect. A scrotum is part of the male reproductive system, and does not fit in this concept map.

C Incorrect. A zygote is the name of a fertilized egg before it is implanted in the uterus. The zygote is not part of the female reproductive system, because it does not occur without sperm and fertilization.

D Incorrect. Menstruation refers to a woman's monthly cycle and is not an organ like the other items on the concept map.

Diagnostic Teaching Tip: Students who answer this question incorrectly might need to review a female's reproduction system. Give students a picture of a female's reproductive system, and have them label and describe the functions of the different parts.

Question 12 *asks students to identify the location of DNA in a cell.*

A Correct. Chromosomes are located in the nucleus of eukaryotes, such as humans.

B Incorrect. A vacuole is a storage organelle found in plant cells.

C Incorrect. Chloroplasts are photosynthetic organelles found in plants and other photosynthetic organisms. Human cells do not contain chloroplasts.

D Incorrect. Mitochondria are organelles that break down food molecules to release the energy needed for cell functions.

Diagnostic Teaching Tip: Students who answer this question incorrectly may benefit from reviewing the parts and functions of plant and animal cells. Create a table in which students enter the names, functions of cell parts, and types of cells where the part may be found.

Question 13 *asks students to match DNA with its description.*

A Correct. Deoxyribonucleic acid, or DNA, is the genetic material of living organisms. It is located in the chromosomes of each cell. The chromosomes of a eukaryotic organism are found in the nucleus. Prokaryotes have a single circular chromosome that is found in the cytoplasm of the cell. Prokaryotes do not have nuclei.

B Incorrect. This is the name of the cell organelle that produces, packages, and moves proteins and lipids.

C Incorrect. ATP, or adenosine triphosphate, is the energy currency of cells.

D Incorrect. A restriction endonuclease is an enzyme that cuts DNA.

Diagnostic Teaching Tip: Students who answer this question incorrectly might need to review what DNA is. Have students research the components of DNA and make a model showing how the components make up DNA.

Standards Assessment *continued*

Question 14 *asks students to understand inherited traits.*

A Incorrect. A child has all their inherited traits when they are born, but certain traits are only apparent as a child grows.

B Incorrect. Cell differentiation is the process by which cells change to perform different processes. Cell differentiation does not determine traits.

C Correct. Genes determine what traits a child will have. One or more genes can be responsible for determining the same trait.

D Incorrect. A child inherits half of the father's genes and half of the mother's genes. The genes that are passed on are not determined through conscious selection of the parent, so the parent is unable to determine which traits will be inherited.

Diagnostic Teaching Tip: Students who answer this question incorrectly might need to review what traits are inherited and which are environmental. Have students make a chart with a column for inherited traits and a column for non-inherited traits. Have students brainstorm different traits, talents, characteristics that they have and decide if they were inherited from their parents or not.

Question 15 *asks students to distinguish the methods of sexual and asexual reproduction.*

A Incorrect. Potatoes form a tuber, an extension of themselves to create a new plant. This is a type of asexual reproduction.

B Correct. Sexual organisms reproduce when two parents both provide a single sex cell. The cells then combine to create a new organism.

C Incorrect. Strawberries send out runners that create new roots in the ground and create new strawberry plants. This is a type of asexual reproduction.

D Incorrect. Many prokaryotes divide through binary fission, which produces an exact replica of the parent organism. This is an example of asexual reproduction.

Diagnostic Teaching Tip: Students who struggle with this type of question might benefit from reviewing sexual and asexual reproduction. Have students make a chart listing the different ways that organisms can reproduce sexually and asexually.

Standards Assessment

REVIEWING ACADEMIC VOCABULARY

_______ **1.** In the sentence "The placenta performs an important function for the fetus," what does the word *function* mean?
    A position
    B organ
    C purpose
    D result

_______ **2.** Which of the following words best completes the sentence: The uterus is a special _____ in which a baby develops.
    A structure
    B system
    C organism
    D method

_______ **3.** Which of the following words is the closest in meaning to the word *generate*?
    A explain
    B produce
    C identify
    D accomplish

_______ **4.** Which of the following sets of words best completes the following sentence: "In _____, offspring receive traits from both of their parents."
    A asexual reproduction
    B sexual function
    C sexual selection
    D sexual reproduction

_______ **5.** What is the noun used to describe "the process in which cells change to carry out specialized functions"?
    A production
    B fabrication
    C distinction
    D differentiation

REVIEWING CONCEPTS

_______ **6.** Which of the following best describes the function of the placenta?
A The placenta provides oxygen and nutrients to the fetus.
B The placenta is the organ that holds the fetus as it grows and develops.
C The placenta is a thin membrane that fills with fluid to protect the fetus.
D The placenta sends nutrients to the mother's body.

_______ **7.** Which of the following can cause a woman to become pregnant?
A A man ejaculates sperm in or near the woman's vagina.
B The woman experiences a menstrual cycle.
C A man ejaculates his sperm during puberty.
D The woman ovulates until middle age.

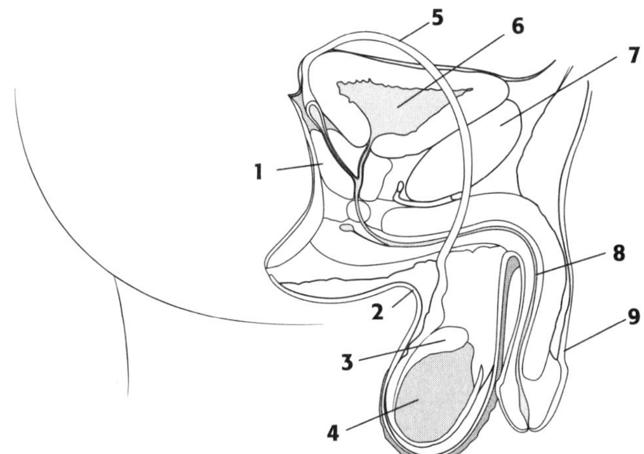

_______ **8.** In the diagram above, which number labels the structure that produces sperm?
A 2
B 4
C 5
D 9

_______ **9.** In humans, what has to occur for an egg to be considered fertilized?
A A sperm has to penetrate the outer coating of the egg.
B An egg has to be present when sperm enter the woman.
C A few hundred sperm have to cover the outer coating of the egg.
D The sperm nucleus and the egg nucleus must join.

_______ **10.** Why does a child usually have physical characteristics from both parents?
A The child inherited genes from each parent.
B The child's genes are mixed while the child is in the uterus.
C The child's cells differentiate during development.
D The environment of the child is the same as that of the parents.

Standards Assessment *continued*

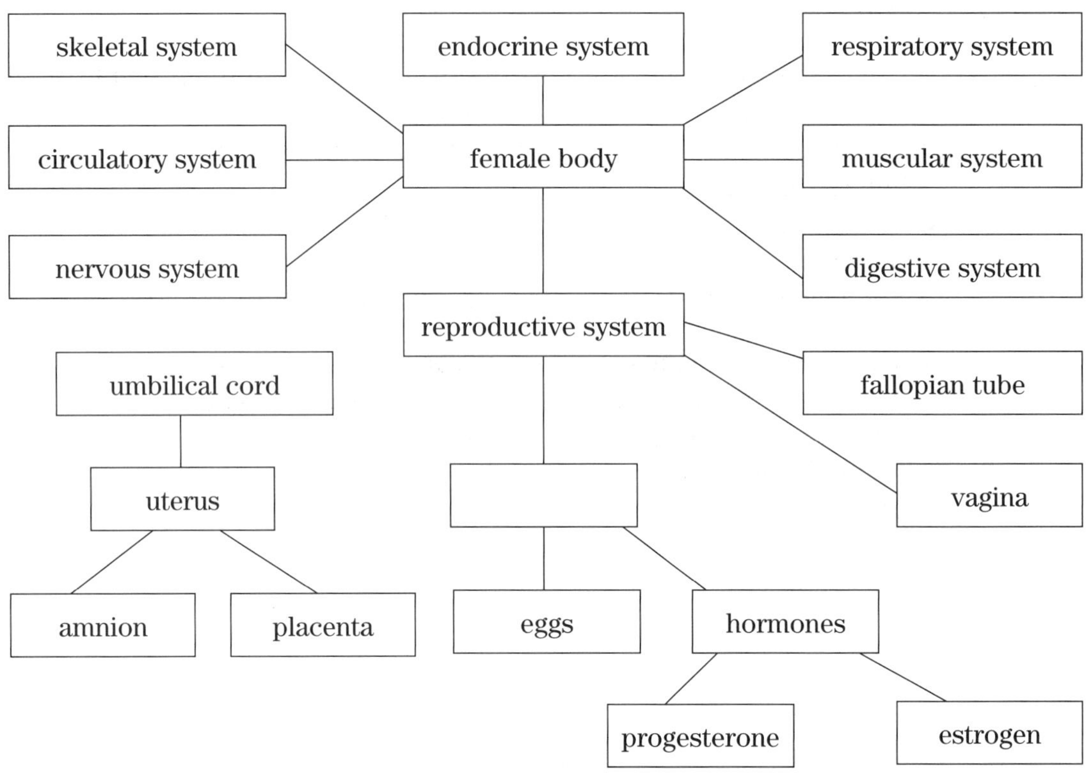

______ **11.** Which of the following best fits into the blank space in the concept map above?
- **A** ovary
- **B** scrotum
- **C** zygote
- **D** menstruation

REVIEWING PRIOR LEARNING

______ **12.** In a human cell, where are chromosomes located?
- **A** nucleus
- **B** vacuole
- **C** chloroplasts
- **D** mitochondria

______ **13.** Which of the following is the genetic material of all organisms?
- **A** deoxyribonucleic acid
- **B** endoplasmic reticulum
- **C** adenosine triphosphate
- **D** restriction endonuclease

Standards Assessment *continued*

_______ **14.** What determines an inherited trait?
    A Traits are inherited as a child grows older.
    B Cell differentiation determines inherited traits.
    C Inherited traits are determined by one or more genes.
    D A child's parents decide what traits will be inherited.

_______ **15.** Which of these describes a reproduction method of a sexual organism?
    A forming a tuber
    B fusing of sex cells from two parents
    C producing runners
    D division through binary fission

　　　　　　　　　　　　　　　　　　　　DATASHEET

Dance of the Chromosomes

Teacher Notes

Students will be modeling what happens in a cell that has one pair of chromosomes as it undergoes meiosis (covers standard 7.2.b). This activity will review meiosis as it prepares students for the discussion of gamete formation in this chapter. Encourage students to bring any instrumental dance music that they enjoy so that the music can help them feel comfortable. Work with groups during the dance creation process to ensure that all groups are depicting the division of diploid cells to create haploid cells.

MATERIALS

For each group

No materials are needed, but an open area is needed for groups to practice and then demonstrate their dances.

Explore Activity) **DATASHEET A**

Dance of the Chromosomes

During meiosis, cells that have two copies of each chromosome divide to form sex cells that have one copy of each chromosome. This is an important process in human reproduction. Females produce sex cells called eggs. Males produce sex cells called sperm. Create a dance that illustrates the steps involved in the formation of egg and sperm cells.

PROCEDURE

1. Review the process of meiosis as presented in Chapter 6, Heredity.

2. Work in a group with three other students.

3. Begin with a diploid cell that contains one pair of chromosomes.
   - A diploid cell has two copies of each chromosome.
   - Each group member will be one chromosome.
   - Initially the cell will have two chromosomes (people).
   - When the DNA of the cell replicates before meiosis, two people will join the first pair.
   - You should now have two homologous chromosomes each with a sister chromatid (or four people in two pairs).
   - The people who are sister chromatids of one chromosome might hold hands to show what they are.

4. Create a dance that illustrates the behavior of these chromosomes during the formation of haploid sex cells. Your dance should show the following steps.
   - Start with one pair of homologous chromosomes, each with two sister chromatids.
   - Homologous chromosomes line up along the equator of the cell.
   - The homologous chromosomes separate from their pair and move to opposite ends of the cell and the cell divides.
   - Paired chromatids are still together. Each cell should have one member of the homologous pair of chromosomes.
   - Chromosomes line up along the equator of each cell.
   - The chromatids pull apart and move to opposite ends of the cell, and each cell divides.
   - Four haploid cells should be formed.
   - Haploid cells have one copy of each chromosome.

5. Perform your dance for your classmates.

ANALYSIS

6. When fertilization occurs, sperm and egg each contribute genetic material to the fertilized egg. Why is it important that sperm and eggs have only one copy of each chromosome?

____

____

7. Because of meiosis, sperm and egg cells are haploid. When a sperm fertilizes an egg, how many copies of each chromosome does the new cell have?

____

____

 DATASHEET B

Dance of the Chromosomes

During meiosis, cells that have two copies of each chromosome divide to form sex cells that have one copy of each chromosome. This is an important process in human reproduction. Females produce sex cells called eggs. Males produce sex cells called sperm. Create a dance that illustrates the steps involved in the formation of egg and sperm cells.

PROCEDURE

1. Review the process of meiosis as presented in Chapter 6, Heredity.

2. Work in a group with three other students.

3. Begin with a diploid cell that contains one pair of chromosomes. A diploid cell has two copies of each chromosome. To each group member assign the role of one chromosome or other cell part.

4. Create a dance that illustrates the behavior of these chromosomes during the formation of haploid sex cells. Haploid cells have one copy of each chromosome.

5. Perform your dance for your classmates.

ANALYSIS

6. Why is it important that sperm and eggs have only one copy of each chromosome?

____

____

____

____

____

7. When a sperm fertilizes an egg, how many copies of each chromosome does the new cell have?

____

____

____

____

____

Dance of the Chromosomes

During meiosis, cells that have two copies of each chromosome divide to form sex cells that have one copy of each chromosome. This is an important process in human reproduction. Females produce sex cells called eggs. Males produce sex cells called sperm. Create a dance that illustrates the steps involved in the formation of egg and sperm cells.

PROCEDURE

1. Review the process of meiosis as presented in Chapter 6, Heredity.

2. Work in a group with three other students.

3. Begin with a diploid cell that contains one pair of chromosomes. A diploid cell has two copies of each chromosome. To each group member assign the role of one chromosome or other cell part.

4. Create a dance that illustrates the behavior of these chromosomes during the formation of haploid sex cells. Haploid cells have one copy of each chromosome.

5. Perform your dance for your classmates.

ANALYSIS

6. Explain why it is important that sperm and eggs have only one copy of each chromosome. What would happen if sperm and egg cells were produced by mitosis instead of meiosis?

____

____

____

7. Explain how the diploid number of chromosomes in an organism is restored when a sperm fertilizes an egg.

____

____

____

DATASHEET

Modeling Inheritance

Teacher Notes

Before completing this lab, review the inheritance of genetic material in humans using **Figure 3** in this section. Put the following chart on the board. This lab helps students understand genetic inheritance (covers standard 7.2.b).

	Homozygous Dominant	Heterozygous Dominant	Homozygous Recessive
Fur color	FF- Blue Fur	Ff- Blue Fur	ff- Red Fur
Number of eyes	EE- 5 eyes	Ee- 5 eyes	ee- 3 eyes
Number of antennae	AA- 4 antennae	Aa- 4 antennae	aa- 6 antennae

MATERIALS

For each student

- coin (1)

Quick Lab

Modeling Inheritance

During sexual reproduction, parents contribute genetic information to offspring. In this activity, you will model inheritance, the passing on of genetic information from parent to offspring.

PROCEDURE

1. A trait is a form of a genetic characteristic.
- For example, blue fur is a form of the characteristic for fur color.
- For most characteristics, there is more than one trait that a parent may pass along to its offspring during sexual reproduction.
- The gene for each trait is called an *allele*.
- Alleles are represented by letters: capital letters for dominant alleles and lowercase letters for recessive alleles.
- Remember diploid organisms have two copies of each gene for a trait.
- Dominant traits are expressed even if only one copy is the dominant allele. Recessive traits are expressed only if both copies are recessive.
- For each characteristic, each parent gives one allele to each offspring.
- By knowing an offspring's two alleles for a characteristic, you can determine what the offspring looks like for that characteristic.

2. For each characteristic, flip a **coin** to pick which allele each parent will pass to the offspring.
- Heads represents a dominant allele (capital letter).
- Tails represents a recessive allele (lowercase letter).

Parent Alleles				
Characteristic	**Mother**		**Father**	
Fur color	F	f	F	f
Number of eyes	E	e	E	e
Number of antennae	A	a	A	a

3. Record the results of each toss in the data table above by circling the allele that each parent contributed.

4. Write the alleles that the offspring receives for each trait, such as FF, Ee, and aa.

____

____

____

Modeling Inheritance *continued*

5. Use the chart that your teacher drew to decide what the offspring in your model looks like.
  - For example, offspring with FF or Ff would have blue fur.
  - Sketch the offspring you produced.
  - Show your work below.

6. How much of the offspring's genetic material did each parent contribute?

____

____

7. Each parent has blue fur, five eyes, and four antennae.
  - What color fur does your offspring have?

____

  - How many eyes does your offspring have?

____

  - How many antennae does your offspring have?

____

  - Does the offspring look like the parent? Why or why not?

____

____

____

____

Modeling Inheritance

During sexual reproduction, parents contribute genetic information to offspring. In this activity, you will model inheritance, the passing on of genetic information from parent to offspring.

PROCEDURE

1. Copy the table shown below. A trait is a form of a genetic characteristic. For example, blue fur is a form of fur color. For most characteristics, there is more than one trait that a parent may pass along to its offspring during sexual reproduction. The gene for each trait is called an allele. Alleles are represented by letters: capital letters for dominant alleles and lowercase letters for recessive alleles. For each characteristic, each parent gives one allele to each offspring. By knowing an offspring's two alleles for a characteristic, you can determine what the offspring looks like for that characteristic.

Parent Alleles						
Characteristic	**Mother**			**Father**		
Fur color	F		f	F		f
Number of eyes	E		e	E		e
Number of antennae	A		a	A		a

2. For each characteristic, flip a **coin** to pick which allele each parent will pass to the offspring. Heads represents a dominant allele (capital letter). Tails represents a recessive allele (lower case letter).

3. Record your results in your data table by circling the allele that each parent contributed.

4. Below your table, write the alleles that the offspring receives, such as FF, Ee, and aa.

| Modeling Inheritance *continued*

5. Use the chart that your teacher drew to decide what the offspring in your model looks like. Sketch the offspring.

6. What did each parent contribute to the offspring?

____

____

____

____

____

7. Each parent has blue fur, five eyes, and four antennae. Does the offspring look like the parent? Why or why not?

____

____

____

____

____

Modeling Inheritance

During sexual reproduction, parents contribute genetic information to offspring. In this activity, you will model inheritance, the passing on of genetic information from parent to offspring.

PROCEDURE

1. A trait is a form of a genetic characteristic. For example, blue fur is a form of fur color. For most characteristics, there is more than one trait that a parent may pass along to its offspring during sexual reproduction. The gene for each trait is called an *allele*. Alleles are represented by letters: capital letters for dominant alleles and lowercase letters for recessive alleles. For each characteristic, each parent gives one allele to each offspring. By knowing an offspring's two alleles for a characteristic, you can determine what the offspring looks like for that characteristic.

2. For each characteristic, flip a **coin** to pick which allele each parent will pass to the offspring. Heads represents a dominant allele (capital letter). Tails represents a recessive allele (lowercase letter).

Parent Alleles				
Characteristic	**Mother**		**Father**	
Fur color	F	f	F	f
Number of eyes	E	e	E	e
Number of antennae	A	a	A	a

____

____

____

3. Record your results in the data table above by circling the allele that each parent contributed.

4. Below your table, write the alleles that the offspring receives, such as FF, Ee, and aa.

| Modeling Inheritance *continued*

---

5. Use the chart that your teacher drew to decide what the offspring in your model looks like. Sketch the offspring.

6. Describe the genetic contribution of each parent to the offspring.

____

____

7. Each parent has blue fur, five eyes, and four antennae. Does the offspring look like the parent? Explain how it is possible for offspring to express traits different from those of the parents.

____

____

____

____

____

____

DATASHEET

Development Timeline

Teacher Notes

This lab helps students understand a fetus's development during pregnancy (covers standards 7.1.f, 7.5.e, and 7.7.d). Make sure students tape the two pieces of paper side-to-side in the vertical position so that they have enough length to list all of the weeks on their timeline and have enough width to draw the lines as long as necessary. You may want to prepare the paper for the students before they begin the lab.

MATERIALS

For each student

- paper, 2 pieces
- tape

 DATASHEET A

Development Timeline

PROCEDURE

Use the timeline in **Figure 3** in your book and the text in this section to show the developing human's size increase during pregnancy.

1. Take **two pieces of paper** and **tape** them together.

2. Along the long edge of the paper, write the week numbers as shown in **Figure 3** in your book.

3. Above each appropriate week number, draw a line that represents the size of the embryo or fetus at that week.
   - Use Figure 3 and the text in this section to list the major developments that occur at each stage.

4. Look at your text. When does the umbilical cord form?

____

____

5. The placenta is a two-way exchange organ where blood between the embryo/ fetus and the mother come in close contact.
   - What are the two purposes of the placenta during fetal development?

____

____

6. Look at your text. When do the embryo's blood cells begin to differentiate, or become specialized?

____

____

7. Look at the lines you drew to represent the size of the embryo or fetus at various stages of development. Between what weeks does the largest growth spurt occur?

____

____

Quick Lab

Development Timeline

PROCEDURE

Use the timeline in **Figure 3** in your book and the text in this section to show the developing human's size increase during pregnancy.

1. Take **two pieces of paper** and **tape** them together.

2. Along the long edge of the paper, write the week numbers as shown in **Figure 3** in your book.

3. Above each appropriate week number, draw a line that represents the size of the embryo or fetus at that week. List the major developments that occur at each stage.

4. When does the umbilical cord form?

____

5. What is the placenta's role in fetal development?

____

____

____

____

6. When do the embryo's blood cells begin to differentiate, or become specialized?

____

7. Between what weeks does the largest growth spurt occur?

____

Quick Lab)   **DATASHEET C**

Development Timeline

PROCEDURE

Use the timeline in **Figure 3** in your book and the text in this section to show the developing human's size increase during pregnancy.

1. Take **two pieces of paper** and **tape** them together.

2. Along the long edge of the paper, write the week numbers as shown in **Figure 3**.

3. Above each appropriate week number, draw a line that represents the size of the embryo or fetus at that week. List the major developments that occur at each stage.

4. When does the umbilical cord form? Explain the role of the umbilical cord.

____

____

5. Explain the role of the placenta during fetal development.

____

____

6. When do the embryo's blood cells begin to differentiate?

____

____

7. Between what weeks does the largest growth spurt occur?

____

____

Skills Practice Lab

DATASHEET

It's a Comfy, Safe World!

Teacher Notes

This lab demonstrates the relationship between the structure
and function of reproductive organs. Using models, this lab
also illustrates the role and importance of the umbilicus and
umbilical cord during pregnancy (covers standards 7.5.e, 7.7.c,
and 7.7.d).

Christine Erskine
Brier Elementary
School
Fremont, California

TIME REQUIRED

Two 45-minute class periods

LAB RATINGS

Easy ← 1　2　3　4 → Hard

Teacher Prep–2
Student Set-Up–1
Concept Level–1
Clean Up–3

MATERIALS

This lab may require some large plastic bags, a meterstick, and various other
materials depending on the students' designs. Soft-boiled eggs will simplify the
cleanup. Students may want to wear gloves.

SAFETY CAUTION

Remind students to review all safety cautions and icons before beginning this lab
activity. Students should wash their hands after handling the eggs.

PREPARATION NOTES

Students will be dropping their models, so this lab should be done over a large
plastic sheet. You may want to do this lab outside, because it may be quite messy.
Students can modify their models in any way they feel will improve the protection.

DATASHEET A

It's a Comfy, Safe World!

Before hatching, baby birds live inside a hard, protective shell until the baby has used up all of the food supply. Before birth, most mammal babies develop within their mother's uterus, in which they are surrounded by fluid and connected to a placenta. Before human babies are born, they lead a comfy life. By the seventh month of development, they suck their thumb, blink their eyes, and perhaps even dream.

OBJECTIVES

Construct a model of a human uterus protecting a fetus.

Compare the protection that a bird's egg gives a developing baby bird with the protection that a human uterus gives a fetus.

MATERIALS

- computer (optional)
- cotton, soft fabric, or other soft materials
- eggs, soft-boiled and in the shell (2 to 4)
- eggs, soft-boiled and peeled (3 or 4)
- gloves, protective
- mineral oil, cooking oil, syrup, or other thick liquid
- plastic bags, sealable
- water

SAFETY INFORMATION 

Using Scientific Methods

ASK A QUESTION

1. Inside which structure is a developing organism better protected from bumps and blows: the uterus of a placental mammal or the egg of a bird?

FORM A HYPOTHESIS

2. A placental mammal's uterus protects a developing organism from bumps and blows better than a bird's egg does.

It's a Comfy, Safe World! *continued*

TEST THE HYPOTHESIS

3. Test the above hypothesis by building a model uterus.
- Brainstorm ways to construct your model of a mammalian uterus.
- Remember that the fetus floats in amniotic fluid within the uterus.
- Use the materials provided by your teacher to build your model.
- A peeled, soft-boiled egg will represent the fetus inside your model uterus.

4. Use **Table 1** below to record your data.
- Test your model by dropping it from a height of 1 m.
- Examine the egg for damage. Record the results of your tests in Table 1 under the "Original Model" column.

Table 1	First Test of Model Uterus
Original model	**Modified model**

5. If the egg is damaged during your test, modify your model as necessary to provide more protection for the egg.
- Test your modified model by using another peeled, soft-boiled egg.
- Drop from a height of 1 m.
- Examine the egg for damage. Record your results under the "Modified Model" column in Table 1.
- If the second egg is damaged, repeat the above steps until you are satisfied with the protection provided for the egg.

6. When you have completed the model's design, obtain another peeled, soft-boiled egg and a soft-boiled egg in the shell.
- The peeled, soft-boiled egg still represents the fetus in your model uterus.
- The egg in the shell represents the baby bird inside the egg.

7. Use **Table 2** below to record your data.
- Test only the peeled egg inside the model. Drop from a height of 1 m.
- Then, test the egg in the shell as is. Drop from a height of 1 m.
- Examine the eggs for damage. Record the results in your data table.

Table 2	Final Test of Model Uterus
	Test results
Model	
Egg in shell	

It's a Comfy, Safe World! *continued*

ANALYZE THE RESULTS

8. Explaining Events What happened to the egg inside your model? Was it damaged by your tests?

____

____

• What happened to the egg in the shell? Was it damaged by the test?

____

____

9. Analyzing Results Was the egg protected inside your model damaged by the drop test?

____

____

• What parts of your model protected the "fetus" from damage?

____

____

DRAW CONCLUSIONS

10. Evaluating Data Review your hypothesis. Was the egg in your model uterus more protected from damage than the egg in the shell?

____

____

• Did your data support the hypothesis? Why or why not?

____

____

11. Evaluating Models Are parts of a uterus missing from your model?

____

____

• How could you make your model even more like a uterus?

____

____

____

It's a Comfy, Safe World! *continued*

BIG IDEA QUESTION

12. Analyzing Relationships How are a human placenta and umbilical cord specialized to protect and provide for a developing human before he or she is born?

____

____

____

____

____

____

APPLYING YOUR DATA

Use the Internet or the library to find information about the development of monotremes, such as the echidna or the platypus, and the development of marsupials, such as the koala or the kangaroo. Then, using what you have learned in this lab, compare the development of placental mammals with the development of marsupials and monotremes.

____

____

____

____

____

____

____

It's a Comfy, Safe World!

Before hatching, baby birds live inside a hard, protective shell until the baby has used up all of the food supply. Before birth, most mammal babies develop within their mother's uterus, in which they are surrounded by fluid and connected to a placenta. Before human babies are born, they lead a comfy life. By the seventh month of development, they suck their thumb, blink their eyes, and perhaps even dream.

OBJECTIVES

Construct a model of a human uterus protecting a fetus.

Compare the protection that a bird's egg gives a developing baby bird with the protection that a human uterus gives a fetus.

MATERIALS

- computer (optional)
- cotton, soft fabric, or other soft materials
- eggs, soft-boiled and in the shell (2 to 4)
- eggs, soft-boiled and peeled (3 or 4)
- gloves, protective
- mineral oil, cooking oil, syrup, or other thick liquid
- plastic bags, sealable
- water

SAFETY INFORMATION 

Using Scientific Methods

ASK A QUESTION

1. Inside which structure is a developing organism better protected from bumps and blows: the uterus of a placental mammal or the egg of a bird?

FORM A HYPOTHESIS

2. A placental mammal's uterus protects a developing organism from bumps and blows better than a bird's egg does.

TEST THE HYPOTHESIS

3. Brainstorm ways to construct and test your model of a mammalian uterus. Then, use the materials provided by your teacher to build your model. A peeled, soft-boiled egg will represent the fetus inside your model uterus.

It's a Comfy, Safe World! *continued*

4. Make a data table similar to **Table 1** below. Test your model, examine the egg for damage, and record your results.

Table 1	First Test of Model Uterus
Original model	**Modified model**

5. Modify your model as necessary; test this modified model by using another peeled, soft-boiled egg; and record your results.

6. When you have completed the model's design, obtain another peeled, soft-boiled egg and a soft-boiled egg in the shell. The egg in the shell represents the baby bird inside the egg.

7. Make a data table similar to **Table 2** below. Test only the peeled egg inside the model. Then, test the egg in the shell as is. Examine the eggs for damage. Record the results in your data table.

Table 2	Final Test of Model Uterus
	Test results
Model	
Egg in shell	

ANALYZE THE RESULTS

8. Explaining Events Explain how the test results for the model differ from the test results for the egg in a shell.

____

____

____

____

____

____

____

____

It's a Comfy, Safe World! *continued*

9. Analyzing Results What modification to your model protected the "fetus" most effectively?

____

____

____

____

____

DRAW CONCLUSIONS

10. Evaluating Data Review your hypothesis. Did your data support your hypothesis? Why or why not?

____

____

____

____

____

11. Evaluating Models What modifications to your model might make it more like a uterus?

____

____

____

____

____

____

It's a Comfy, Safe World! *continued*

BIG IDEA QUESTION

12. Analyzing Relationships How is a human placenta and umbilical cord specialized to protect and provide for a developing human before he or she is born?

____

____

____

____

____

____

____

____

APPLYING YOUR DATA

Use the Internet or the library to find information about the development of monotremes, such as the echidna or the platypus, and the development of marsupials, such as the koala or the kangaroo. Then, using what you have learned in this lab, compare the development of placental mammals with the development of marsupials and monotremes.

____

____

____

____

____

____

____

____

  DATASHEET C

It's a Comfy, Safe World!

Before hatching, baby birds live inside a hard, protective shell until the baby has used up all of the food supply. Before birth, most mammal babies develop within their mother's uterus, in which they are surrounded by fluid and connected to a placenta. Before human babies are born, they lead a comfy life. By the seventh month of development, they suck their thumb, blink their eyes, and perhaps even dream.

OBJECTIVES

Construct a model of a human uterus protecting a fetus.

Compare the protection that a bird's egg gives a developing baby bird with the protection that a human uterus gives a fetus.

MATERIAL

- computer (optional)
- cotton, soft fabric, or other soft materials
- eggs, soft-boiled and in the shell (2 to 4)
- eggs, soft-boiled and peeled (3 or 4)
- gloves, protective
- mineral oil, cooking oil, syrup, or other thick liquid
- plastic bags, sealable
- water

SAFETY INFORMATION 

ASK A QUESTION

1. Inside which structure is a developing organism better protected from bumps and blows: the uterus of a placental mammal or the egg of a bird?

FORM A HYPOTHESIS

2. Formulate a hypothesis that answers the question above.

____

____

It's a Comfy, Safe World! *continued*

TEST THE HYPOTHESIS

3. Brainstorm ways to construct and test a model of a mammalian uterus. Then, use the materials provided by your teacher to build your model. A peeled, soft-boiled egg will represent the fetus inside your model uterus.

4. Determine how you will test the protection provided to the "fetus" by your model. Make a data table to record the results of your tests.

5. Modify your model as necessary; test this modified model by using another peeled, soft-boiled egg; and record your results.

6. When you have completed the model's design, obtain another peeled, soft-boiled egg and a soft-boiled egg in the shell. The egg in the shell represents the baby bird inside the egg.

7. Test only the peeled egg inside the model. Then, test the egg in the shell as is. Examine the eggs for damage. Record the results in another data table.

ANALYZE THE RESULTS

8. Explaining Events Compare the test results for your model uterus to the test results for the egg in a shell.

____

____

____

____

9. Analyzing Results Describe the model modifications that protected the "fetus" most effectively.

____

____

____

____

____

It's a Comfy, Safe World! *continued*

DRAW CONCLUSIONS

10. Evaluating Data Did your data support your hypothesis? Why or why not?

____

____

____

11. Evaluating Models Describe modifications you could make to your model to make it more like a uterus.

____

____

____

____

BIG IDEA QUESTION

12. Analyzing Relationships Explain how a human placenta and umbilical cord are specialized to protect and provide for a developing human before he or she is born.

____

____

____

____

____

APPLYING YOUR DATA

Use the Internet or the library to find information about the development of monotremes, such as the echidna or the platypus, and the development of marsupials, such as the koala or the kangaroo. Then, using what you have learned in this lab, compare the development of placental mammals with the development of marsupials and monotremes in a short report on a separate sheet of paper.

DATASHEET

Using Print Resources for Research

Teacher Notes

After students have made a list of key words, tell them that most research is completed for the betterment of people. Explain that, to this end, most researchers share their ideas with other researchers. Encourage students to share their key word lists with each other.

Science Skills Activity)   **DATASHEET**

Using Print Resources for Research

INVESTIGATION AND EXPERIMENTATION

7.7.b Use a variety of print and electronic resources (including the World Wide Web) to collect information and evidence as part of a research project.

TUTORIAL

Print resources that you might use in a research project include newspapers, magazines, journals, encyclopedias, and other books. Reading is one way to get information from a print resource. But there are several other ways to quickly get information from a print resource. The following tips are ways to help you quickly find what you are looking for.

Procedure

1. **List Your Keywords** After you have gathered your print resources, make a list of keywords that are important to your research project.

2. **Find the Main Idea** Often, the main idea of the paragraph is stated in the first sentence. And the last sentence restates that idea.

3. **Important Headings and Illustrations** When you find a page that appears to have useful information, read the headings on that page to see if they relate to your topic. Then, look at any pictures, diagrams, charts, or maps on the page to see if they relate to your topic. Be sure to read the captions.

4. **Scan the Resource** Read only a few words here and there. Scan a passage in order to find keywords. By scanning, you can decide which parts of the text you should concentrate on.

5. **Skim the Resource** Read only a sentence or two. Look for sentences that look especially important. Skim a passage to get a general idea of what it is about or to determine if you want to read some parts more carefully.

YOU TRY IT!

Some scientists study the reproduction and development of animals to understand human reproduction and development. Use chapter 14 in this textbook and the procedure above to research how and why animals reproduce.

Procedure

1. Before you begin, what keywords will be helpful to search for?

____

____

____

Using Print Resources for Research *continued*

2. Follow each of the steps outlined above and make notes in a chart similar to the one below.

> Research Topic:
>
> Key Words:
>
> Print Source:
>
> Important Information:

Analysis

3. Describing Why is choosing your key words before you begin your research important?

____

____

____

4. Evaluating Did this print resource include useful information for your research topic?

____

____

____

5. Designing What other print resources would be useful for researching this topic? Newspapers, journals, and magazines often have information about new discoveries. Books usually have explanations of processes and facts that have been accepted by many scientists.

____

____

____

6. Concluding Describe the additional steps that you would need to follow to continue your research on this topic. Where might you find print resources that are useful for your research?

____

____

____

____

Answer Key

Directed Reading A

SECTION: HUMAN REPRODUCTION
1. D
2. B
3. A
4. C
5. C
6. D
7. A
8. B
9. B
10. C
11. C
12. B
13. D
14. A
15. C
16. B
17. C
18. A
19. B
20. B
21. A
22. C
23. B
24. A
25. D
26. C
27. A
28. D
29. C
30. D

SECTION: GROWTH AND DEVELOPMENT
1. B
2. C
3. B
4. B
5. B
6. C
7. D
8. B
9. C
10. A
11. D
12. C
13. B
14. C

15. A
16. B
17. D
18. C
19. B
20. C

Directed Reading B

SECTION: HUMAN REPRODUCTION
1. A
2. F
3. D
4. E
5. C
6. B
7. G
8. H
9. D
10. B
11. A
12. C
13. C
14. A woman begins menstruating at puberty. A woman usually stops menstruating in her late 40s or early 50s.
15. blood and tissue
16. ovulation
17. 28 days
18. within a few days of ovulation
19. identical twins
20. one-third
21. fraternal twins
22. quintuplets
23. C
24. D
25. C
26. B
27. A
28. B
29. C
30. STD-related infertility occurs less often in men than in women.

SECTION: GROWTH AND DEVELOPMENT
1. C
2. B
3. D
4. A

TEACHER RESOURCES

5. A
6. D
7. B
8. A
9. B
10. E
11. C
12. D
13. C
14. D
15. A
16. D
17. puberty
18. Answers may vary. Sample answers: His body becomes more muscular; his voice becomes deeper; body and facial hair appear.
19. Answers may vary. Sample answers: The amount of fat in the hips and thighs increases; the breasts enlarge; body hair appears; menstruation begins.
20. Answers may vary. Sample answers: Muscles lose their flexibility; eyesight deteriorates; body fat increases; there is some hair loss.
21. Answers may vary. Sample answers: Hair may turn gray; athletic abilities decline; skin may wrinkle.

Vocabulary and Section Summary A

SECTION: HUMAN REPRODUCTION

1. testes: the primary male reproductive organs, which produce sperm and testosterone (singular, testis)
2. penis: the male organ that transfers sperm to a female and that carries urine out of the body
3. ovary: in the female reproductive system of animals, an organ that produces eggs
4. uterus: in female placental mammals, the hollow, muscular organ in which an embryo embeds itself and develops into a fetus
5. vagina: the female reproductive organ that connects the outside of the body to the uterus

SECTION: GROWTH AND DEVELOPMENT

1. embryo: in humans, a developing individual, from fertilization through the 10th week of pregnancy
2. placenta: the partly fetal and partly maternal organ by which materials are exchanged between a fetus and the mother
3. pregnancy: in medical practice, the period of time between the first day of a woman's last menstrual period and the delivery of her baby (about 280 days, or 40 weeks)
4. umbilical cord: the ropelike structure through which blood vessels pass and by which a developing mammal is connected to the placenta
5. fetus: a developing human from the end of the 10th week of pregnancy until birth

Vocabulary and Section Summary B

SECTION: HUMAN REPRODUCTION

1. penis
2. ovary
3. testes
4. vagina
5. uterus

SECTION: GROWTH AND DEVELOPMENT

1. placenta
2. pregnancy
3. embryo
4. fetus
5. umbilical cord

S	Y	G	Q	E	V	Y	T	Z	A	H	Z	Y	S
F	D	P	L	V	A	D	K	S	T	Q	F	E	E
A	T	H	E	Z	J	K	J	E	Y	Z	E	H	A
J	H	A	T	Q	P	L	A	C	E	N	T	A	H
Z	Q	H	Y	Z	C	Z	Y	D	J	E	U	F	Y
Q	J	E	E	E	H	S	Q	X	Q	F	S	D	T
U	M	B	I	L	I	C	A	L	C	O	R	D	X
Y	V	A	C	A	H	T	H	D	J	Y	T	J	D
S	Y	E	Q	K	D	Z	T	E	H	R	D	Q	Y
G	D	X	S	J	A	V	S	V	Z	B	T	J	S
E	Q	T	K	Q	K	E	A	Y	Q	M	H	Q	H
A	D	H	E	J	D	T	H	G	V	E	V	A	T
Y	C	N	A	N	G	E	R	P	T	Y	J	S	Y
Q	D	C	S	F	Z	A	X	E	D	Z	Q	T	A

Reinforcement

EARLY HUMAN DEVELOPMENT

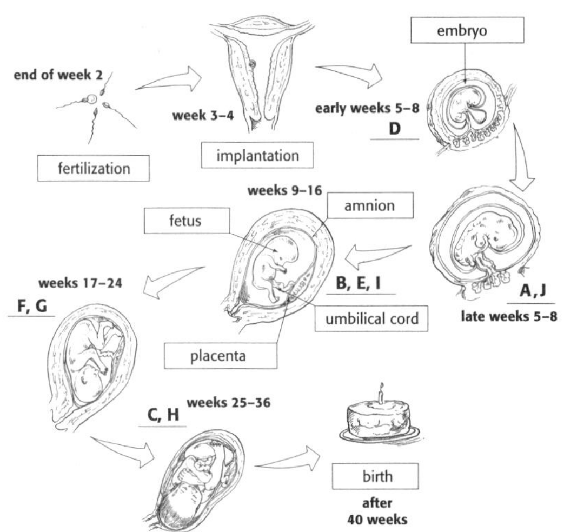

Physical/ Language	Emotional	Social
Tied shoelaces	Pride in accomplishment	Sex appropriate play with trucks

Critical Thinking

1. Answers may vary. Sample answer: Both types of reproduction involve one egg and one sperm cell, which develop into more than one embryo.
2. Answers may vary. Sample answer: The people would be considered identical twins because they are produced from one fertilized egg.
3. Answers may vary. Sample answer: The hatchery would have to provide nourishment for the embryos and a system for removing their waste.
4. Answers may vary. Sample answer: No, because scientists control the reproduction, and they'd make sure the embryos were perfect.
5. Answers may vary. Sample answer: It would be beneficial because it helps create healthy individuals; it would be harmful because people are not given the freedom to determine their own place in society.

SciLinks Activity

Answers may vary. Sample answer: I am almost five years old. Today was a very special day. Today I tied my shoelaces by myself. I tied very slowly and carefully. I hardly got my fingers tangled at all. Mom says she's very proud of me and that I'm a very good boy. She's going to make something special for me for dinner and she said that I can play with my trucks!

Section Review

SECTION: HUMAN REPRODUCTION

1. Sample answer: A woman's vagina is the canal that connects the uterus to the outside world.
2. Answers may vary. In males, testes make sperm and produce testosterone, which help regulate sperm production and the development of male characteristics. The penis is the external male organ that transfers semen into the female's body. In females, ovaries are the organs that make eggs and release estrogen and progesterone, which regulate the release of eggs and the development of female characteristics. The uterus is the organ in which the fertilized egg develops. The vagina is the canal that connects the uterus to the outside world, through which a baby passes when it is born.
3. Sperm and eggs are both sex cells. This means that they each have one copy of each chromosome. Sex cells are made in reproductive organs.
4. One reproductive system problem is infertility. Infertility is the inability to have children. Sexually transmitted diseases (STDs) are a second reproductive system problem. STDs can infect the reproductive system and may cause infertility.
5. An egg cannot be fertilized.
6. Answers may vary. When a single fertilized egg splits in two, each cell contains identical genes. If each cell develops into a separate baby, they will be identical twins. When two separate eggs are fertilized, their genes are different. This results in fraternal twins.
7. $7 \div 1,000 = 0.007$; $0.007 \times 100\% = 0.7\%$; This is less than 1%.

8. Sample answer: The male and female reproductive systems are nourished by nutrients that are carried to them by the circulatory system. The nervous system also gathers information about and coordinates the activities of the reproductive systems. There are many ways that the reproductive system relies on other body systems.

SECTION: GROWTH AND DEVELOPMENT

1. The umbilical cord is a cord that forms during the 5th week of pregnancy and connects the embryo or fetus to the placenta.

2. Sample answer: An embryo is less developed than a fetus is.

3. Sample answer: An embryo begins as a single fertilized egg cell, then develops into a ball of cells that embeds itself in the wall of the uterus. Soon, the amnion, placenta, and umbilical cord form to protect and nourish the embryo. The embryo develops and, after week 10, it is called a fetus. By week 12, all of the major organ systems have begun to form and grow. By week 16, the fetus is moving strongly enough for the mother to feel the movements. From week 12 to week 40, all body systems continue to grow and strengthen. At about week 40, the fetus is ready to be born.

4. Fertilization happens when a sperm's nucleus unites with an egg's nucleus to form a zygote. Implantation takes place when an embryo embeds itself in the wall of the uterus.

5. fertilization to birth, infancy, childhood, adolescence, and adulthood

6. The role of structure a, which is the placenta, is to provide nutrients and oxygen to the fetus or embryo and remove wastes.

7. The covering keeps other sperm from entering the egg and uniting with the nucleus. This is important because it helps maintain the correct chromosome number.

8. Sample answer: Adolescence is the most important stage because that is when humans reach sexual maturity.

Without sexual maturity, the species could not survive.

9. • infancy: 2 y ÷ 80 y = 0.025 = 2.55%
 • childhood (age 2 to age 12): 10 y ÷ 80 y = 0.125 = 12.5%
 • adolescence (age 12 to age 20): 8 y ÷ 80 y = 0.1 = 10%
 • adulthood (age 20 to age 80): 60 y ÷ 80 y = 0.75 = 75%

Chapter Review

1. A

2. An embryo is a developing human, from when the zygote becomes a ball of cells by the third week, through the 10th week of pregnancy. A fetus is a developing human from the 10th week of pregnancy until birth.

3. Testes are male organs that produce sperm. Ovaries are female organs that produce eggs.

4. The uterus is the female organ in which a fertilized egg is embedded and the fetus develops. The vagina is the female organ that connects the uterus to the outside of the body.

5. Fertilization happens when the nucleus of a sperm unites with the nucleus of an egg. Implantation is when the embryo embeds itself in the wall of the uterus.

6. The umbilical cord links the fetus with the placenta. The placenta is the organ that helps the fetus get oxygen and nutrients and helps the fetus get rid of wastes.

7. A

8. C

9. A

10. D

11. The testes produce sperm, and the ovaries produce eggs.

12. The placenta is a specialized organ that allows the fetus to get oxygen and get rid of wastes.

13. infancy, childhood, adolescence, and adulthood; Infancy is the stage from birth to age 2. Childhood lasts from age 2 to puberty. Adolescence is the stage from puberty to adulthood. Adulthood begins at about age 20.

14. Sample answer: Infertility is a problem because it may prevent people from having babies. STDs are a problem because they can damage a person's reproductive system, and they may also cause infertility. Cancer is a problem because it can damage a person's reproductive system and may result in infertility or death.

15. Drawings should resemble the images in Section 1, Figure 1, The Male Reproductive System, and Section 1, Figure 2, The Female Reproductive System.

16. Cells begin to differentiate in weeks 3 and 4 of an embryo's life.

17. Answers may vary. Students should write a main idea or hypothesis about the effect of vitamins on a human fetus. They may describe gathering research or interviewing doctors and mothers about vitamin use during pregnancy.

18. An answer to this exercise can be found at the end of the Teacher Edition.

19. Parents contribute one copy of each gene to their offspring. The sperm carries the chromosomes of the father. The egg contains the chromosomes of the mother.

20. Sample answer: Multiple births happen when a fertilized egg splits or when two or more eggs are fertilized at the same time. It is rare for a fertilized egg to split more than once or for more than two eggs to be fertilized at the same time. So, there are fewer multiple births of higher numbers of babies.

21. Sample answer: A female athlete that exercises a great deal will likely have very little body fat. As a result, she may produce less estrogen and stop menstruating.

22. The name of the cord that connects the fetus to its mother is called the umbilical cord. Its purpose is to connect the fetus to the placenta, which provides nutrients and oxygen to the fetus.

23. Estrogen levels fluctuate, but testosterone levels stay the same throughout the month.

24. Estrogen affects the menstrual cycle.

25. Testosterone levels stay the same because the male reproductive system does not have a cycle like the female reproductive does.

26. Sample answer: yes; I think it might change because estrogen affects the menstrual cycle. During pregnancy, the menstrual cycle stops, so the level of estrogen will level off until the pregnancy is over.

27. About 27 pairs of identical twins; $(2700 \div 100 = 27)$

28. Answers may vary. Sample answer: The placenta is specialized for its function because it allows the blood of the fetus and the blood of the mother to exchange nutrients, waste products, and oxygen without the blood mixing.

Section Quizzes

SECTION: HUMAN REPRODUCTION

1. A
2. A
3. C
4. B
5. D
6. C
7. A
8. C
9. B

SECTION: GROWTH AND DEVELOPMENT

1. C
2. B
3. A
4. D
5. C
6. B
7. C
8. B
9. B
10. B

Chapter Test A

1. D
2. A
3. C
4. B
5. B
6. D

7. D
8. A
9. D
10. B
11. D
12. B
13. A
14. B
15. C
16. C
17. B
18. A
19. A
20. C
21. B
22. adolescence
23. infant
24. young adulthood
25. childhood

Chapter Test B

1. C
2. B
3. A
4. D
5. B
6. B
7. A
8. D
9. C
10. B
11. B
12. C
13. A
14. B
15. C
16. D
17. B
18. E
19. A
20. F
21. C
22. E
23. D
24. F
25. A
26. B
27. C
28. C
29. A
30. B

Chapter Test C

1. embryo
2. sperm
3. pregnancy
4. uterus
5. placenta
6. fetus
7. differentiate
8. C
9. B
10. A
11. D
12. C
13. D
14. Answers may vary. Sample answer: The ovaries produce estrogen and progesterone; these hormones regulate the release of eggs and development of female characteristics.
15. Answers may vary. Sample answer: Sperm are made in the testes and leave the body through the penis.
16. Answers may vary. Sample answer: Identical twins share identical genes and look alike; fraternal twins have different genes and do not look exactly alike.
17. Answers may vary. Sample answer: An embryo develops during the 3rd to 10th week of pregnancy. After week 10, the embryo is called a fetus, and at week 13, the fetus begins to develop human features.
18. Answers may vary. Sample answer: Yes, sexual activity without ejaculation can lead to pregnancy. Even without ejaculation, some sperm may still exit the penis. Therefore, sexual activity that releases only a few sperm can lead to a sperm penetrating and fertilizing the egg.
19. Answers may vary. Sample answer: They need half the number of chromosomes so that when they unite, the offspring has the whole number of chromosomes.
20. Answers may vary. Sample answer: This characteristic is important because it allows the fetus to be squeezed through the narrow pelvic opening without damage to the skull during birth.

21. a. testes, **b.** ovaries, **c.** epididymis, **d.** fallopian tubes, **e.** vas deferens, **f.** uterus

Performance-Based Assessment

4. Adulthood is generally the longest stage.

5. Answers may vary. Sample answer: It was difficult for me to tie my shoes before age 4 or 5 because my brain was not developed enough to remember all the steps involved.

6. Answers may vary. Sample answer: (boys) I wouldn't be able to grow a beard. (girls) I wouldn't be able to have children.

7. Answers may vary. Sample answer: Prior to puberty, your body must develop in a strong and healthy way so that you will be a healthy adult and your body will be prepared to have healthy children.

8. Answers may vary. Sample answer: Humans typically begin to be able to reproduce during adolescence. This is because puberty takes place in adolescence and this is when a person's reproductive system matures.

Explore Activity

DATASHEET A

6. When fertilization occurs, the sperm and egg each contribute one copy of each chromosome to the new fertilized egg cell. If the sperm and egg cells had two copies of each chromosome, the new cell would have twice as many chromosomes as the parent, which is too many.

7. The new cell has two copies of each chromosome.

DATASHEET B

6. When fertilization occurs, the sperm and egg each contribute one copy of each chromosome to the newly fertilized egg cell.

7. The new cell has two copies of each chromosome.

DATASHEET C

6. When fertilization occurs, the sperm and egg each contribute one copy of each chromosome to the newly fertilized egg cell. This ensures that the new cell has the correct number of chromosomes. If sperm and eggs were produced by mitosis, the number of chromosomes in offspring would be double that of the parents.

7. When a haploid sperm fertilizes a haploid egg, the new cell has two copies of each chromosome and therefore is diploid.

Quick Lab: Modeling Inheritance

DATASHEET A

6. Each parent contributed half of the offspring's genetic material.

7. Answers may vary. Students should describe what the offspring looks like and compare the offspring to the parents.

DATASHEET B

6. Each parent contributed half of the offspring's genetic material.

7. Answers may vary. Students should describe what the offspring looks like and compare the offspring to the parents.

DATASHEET C

6. Each parent contributed half of the offspring's genetic material.

7. Answers may vary. Students should describe what the offspring looks like and compare the offspring to the parents. It is possible for the offspring to have traits the parents do not. Each parent is heterozygous for each trait; they have a dominant and recessive allele for each trait. Since each parent can contribute either allele for each trait, for any trait there is a 25% chance the offspring will express the recessive trait

Quick Lab: Development Timeline

DATASHEET A

4. week 5

5. The placenta supplies oxygen and nutrients from the mother's blood and passes wastes from the embryo to the mother's blood so she can excrete them.

6. weeks 3 and 4

7. weeks 17 to 24

DATASHEET B

4. week 5

5. The placenta supplies oxygen and nutrients from the mother's blood and passes wastes from the embryo to the mother's blood so she can excrete them.

6. weeks 3 and 4

7. weeks 17 to 24

DATASHEET C

4. The umbilical cord forms in week 5. It connects the embryo to the placenta. Blood vessels pass through it and connect to the placenta, where nutrients, oxygen, and wastes are exchanged between the embryo or fetus and mother.

5. The placenta supplies oxygen and nutrients from the mother's blood and passes wastes from the embryo or fetus to the mother's blood so she can excrete them.

6. The embryo's cells begin to differentiate in weeks 3 and 4.

7. The biggest growth spurt occurs between weeks 17 and 24.

Chapter Lab

DATASHEET A

8. Answers may vary. An egg floating in a viscous liquid in a plastic bag, wrapped in soft cotton, and placed inside another bag should not be damaged when dropped from a height of 1 m. An egg protected only by a shell should break. In general, the more protected the fetus or egg, the more resistant to damage it is and the less damage it will suffer when dropped from a height of 1 m.

9. Answers may vary according to students' modifications. Most students will observe that the more soft wrapping and/or viscous fluid protecting the egg, the better the egg survived the drop from 1 m.

10. Accept all reasonable answers. Students' answers may vary depending on the data collected.

11. Accept all reasonable answers. Students' answers may vary depending on their model and the data collected.

12. Accept all reasonable answers that describe the function of the placenta and uterus.

Applying Your Data

Students should recognize that all mammals have internal fertilization. Students should include the ideas that monotreme mammals lays eggs with leathery shells and, after the eggs hatch, nurture their young with milk that oozes from pores in the mother's skin. Marsupial mammals give birth to live young that are very underdeveloped; the young move to a pouch and continue their development in the pouch. Placental mammals also give birth to live young, but their young are more developed than those of marsupial mammals. Even so, placental mammals may have to care for their young for several years.

DATASHEET B

8. Answers may vary. An egg floating in a viscous liquid in a plastic bag, wrapped in soft cotton and placed inside another bag should not be damaged when dropped from a height of 1 m. An egg protected only by a shell should break. In general, the more protected the fetus or egg, the more resistant to damage it is and the less damage it will suffer when dropped from a height of 1 m.

9. Answers may vary according to students' modifications. Most students will observe that the more soft wrapping and/or viscous fluid protecting the egg, the better the egg survived the drop from 1 m.

10. Accept all reasonable answers. Students' answers may vary depending on the data collected.

11. Accept all reasonable answers. Students' answers may vary depending on their model and the data collected.

12. Accept all reasonable answers that describe the function of the placenta and uterus.

Applying Your Data

Students should recognize that all mammals have internal fertilization. Students should include the ideas that monotreme mammals lay eggs with leathery shells and, after the eggs hatch, nurture their young with milk that oozes from pores in the mother's skin. Marsupial mammals give birth to live young that are very underdeveloped; the young move to a pouch and continue their development in the pouch. Placental mammals also give birth to live young, but their young are more developed than those of marsupial mammals. Even so, placental mammals may have to care for their young for several years.

DATASHEET C

2. Answers may vary. Sample answer: A placental mammal's uterus protects a developing organism from bumps and blows better than a bird's egg does.

8. Answers may vary. An egg floating in a viscous liquid in a plastic bag, wrapped in soft cotton, and placed inside another bag should not be damaged when dropped from a height of 1 m. An egg protected only by a shell should break. In general, the more protected the fetus or egg, the more resistant to damage it is and the less damage it will suffer when dropped from a height of 1 m.

9. Answers may vary according to students' modifications. Most students will observe that the more soft wrapping and/or viscous fluid protecting the egg, the better the egg survived the drop from 1 m.

10. Accept all reasonable answers. Students' answers may vary depending on their hypotheses and the data collected.

11. Accept all reasonable answers. Students' answers may vary depending on their model and the data collected.

12. Accept all reasonable answers that describe the function of the placenta and uterus.

Applying Your Data

Students should recognize that all mammals have internal fertilization. Students should include the ideas that monotreme mammals lays eggs with leathery shells and, after the eggs hatch, nurture their young with milk that oozes from pores in the mother's skin. Marsupial mammals give birth to live young that are very underdeveloped; the young move to a pouch and continue their development in the pouch. Placental mammals also give birth to live young, but their young are more developed than those of marsupial mammals. Even so, placental mammals may have to care for their young for several years.

Science Skills Activity

DATASHEET

3. Key words help you find what is most important to you when you are reading print resources.

4. Answers may vary. Accept all reasonable answers that describe the student's experience.

5. Answers may vary. Students should list specific print resources.

6. Answers may vary. Students will need to describe where they need to go to find information regarding their research.